RESIDUAL
GOVERNANCE

RESIDUAL GOVERNANCE

HOW
SOUTH
AFRICA
FORETELLS
PLANETARY
FUTURES

GABRIELLE HECHT

DUKE UNIVERSITY PRESS

DURHAM & LONDON 2023

Project Editor: Liz Smith
Designed by Matthew Tauch
Typeset in Untitled Serif and Saira by
Copperline Book Services

Library of Congress Cataloging-in-Publication Data
Names: Hecht, Gabrielle, author.
Title: Residual governance : how South Africa foretells
planetary futures / Gabrielle Hecht.
Description: Durham : Duke University Press, 2023. |
Includes bibliographical references and index.
Identifiers: LCCN 2022061830 (print)
LCCN 2022061831 (ebook)
ISBN 9781478024941 (paperback)
ISBN 9781478020288 (hardcover)
ISBN 9781478027263 (ebook)
ISBN 9781478093688 (ebook other)
Subjects: LCSH: Mineral industries—Environmental aspects—
South Africa. | Mineral industries—Social aspects—South
Africa. | Mines and mineral resources—South Africa. |
Environmental degradation—Social aspects—South Africa.
| Environmental justice—South Africa. | BISAC: SOCIAL
SCIENCE / Ethnic Studies / African Studies | NATURE /
Environmental Conservation & Protection
Classification: LCC HD9506.S62 H44 2023 (print) | LCC
HD9506.S62 (ebook) | DDC 333.8/50968—dc23/eng/20230530
LC record available at https://lccn.loc.gov/2022061830
LC ebook record available at https://lccn.loc.gov/2022061831

Cover art: Potšišo Phasha, *Swimming Upstream (color)*,
2013. © Potšišo Phasha.

Visual epigraphs were composed by
Chaz Maviyane-Davies. http://www.maviyane.com/.

FOR
NOLIZWI

CONTENTS

ABBREVIATIONS   ix

NOTE ON USAGE   xi

INTRODUCTION / **THE RACIAL CONTRACT
IS TECHNOPOLITICAL**   1

1   YOU CAN SEE APARTHEID FROM SPACE   19

2   THE HOLLOW RAND   47

3   THE INSIDE-OUT RAND   85

4   SOUTH AFRICA'S CHERNOBYL?   129

5   LAND MINES   163

CONCLUSION / **LIVING IN A FUTURE WAY
AHEAD OF OUR TIME**   197

ACKNOWLEDGMENTS   209

NOTES   215

BIBLIOGRAPHY   237

INDEX   259

ABBREVIATIONS

AEB	Atomic Energy Board
ALARA	as low as reasonably achievable
AMD	acid mine drainage
ANC	African National Congress
CSIR	Council for Scientific and Industrial Research
CWP	Community Work Program
DA	Democratic Alliance
DEA	Department of Environmental Affairs
DMR	Department of Mineral Resources
DRD	Durban Roodeport Deep
EIA	environmental impact assessment
EMF	Environmental Management Framework
ERGO	East Rand Gold and Uranium Company
FSE	Federation for a Sustainable Environment
GCRO	Gauteng City-Region Observatory
GDARD	Gauteng Department of Agricultural and Rural Development
IAEA	International Atomic Energy Agency
ICRP	International Commission on Radiological Protection
IDC	Industrial Development Corporation
IWQS	Institute for Water Quality Studies
LTG	*Limits to Growth*
NNR	National Nuclear Regulator
PHRAG	Provincial Heritage Resources Authority of Gauteng
RDP	Reconstruction and Development Programme
RMP	Rand Mines Properties
SAF	Strategic Area Framework
SAHRC	South Africa Human Rights Commission
SDF 2040	*Spatial Development Framework 2040*
SERI	Socio-Economic Rights Institute

STS	science and technology studies
WHO	World Health Organization
WNLA	Witwatersrand Native Labor Association
WRC	Water Research Commission
WRDM	West Rand District Municipality

NOTE ON USAGE

GOVERNMENT DEPARTMENTS in South Africa often change names under new administrations. For example, the Department of Minerals and Energy was split into two in 2009: Mineral Resources and Energy. In 2019, these were reunited as the Department of Mineral Resources and Energy. The Department of Environmental Affairs and Tourism was similarly split in two in 2009. In 2019, the Department of Environmental Affairs merged with parts of the Department of Agriculture, Forestry, and Fisheries to become the Department of Environment, Forestry, and Fisheries. For the most part, the narrative uses the nomenclature contemporaneous to the year it discusses.

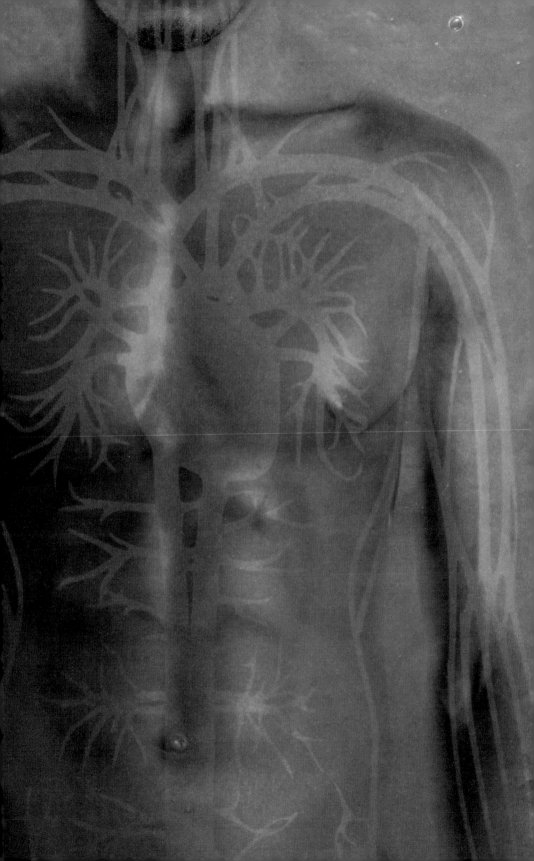

INTRODUCTION
THE RACIAL CONTRACT IS TECHNOPOLITICAL

ON JUNE 16, 1976, Mrs. Sithole, a teacher in Soweto, watched outside her school as thousands of children marched toward Orlando stadium to protest the latest indignity proclaimed by the apartheid regime.[1] The government had revised its Bantu education policy to mandate that all secondary schools use both Afrikaans and English as their language of instruction. Outraged Black teachers—many of whom didn't speak Afrikaans—petitioned the Department of Bantu Education to reconsider. Officials wouldn't budge. "I have not consulted the African people on the language issue," retorted the deputy minister in charge, "and I'm not going to. An African might find that 'the big boss' spoke only Afrikaans or spoke only English. It would be to his advantage to know both languages."[2] Black teachers were incensed. "Why should we in the urban areas have Afrikaans—a language spoken nowhere else in the world and which is still in a raw state of development . . . —pushed down our throats?"[3]

Equally furious, the burgeoning youth and student movement saw the language mandate as the latest in a long line of insults and injuries inflicted upon their schooling since the passage of the 1953 Bantu Education Act. The state spent 42 rands per Black schoolchild, one-fifteenth of the 644 rands it spent per white schoolchild. By 1967, student-teacher ratios in Black classrooms had plummeted to 58:1. Being forced to learn mathematics in Afrikaans—a language that didn't even have a full-fledged scientific vocabulary—was the last straw. School had become untenable. Nourished by the rising Black Consciousness movement, student organizations began to coordinate protests. In May 1976, some six hundred students at Phefeni Junior Secondary School went on strike against the language mandate. More followed at other schools. Building on this momentum, the student movement organized a mass demonstration for June 16.[4]

The march began peacefully. "Yes!" Mrs. Sithole later recalled thinking, "This is now serious." She noticed policemen standing at a street corner. But surely, she thought, they wouldn't hurt the children, who streamed along by the "thousands, all in school uniform." After all, "they had nothing, they were not aggressive. They just had placards saying 'Away with Afrikaans.'" Police vans appeared. The children refused to return to school. And then "the police just opened fire. There were little children coming. . . . I remember their uniform was green and grey. . . . All hell broke loose here. And the little boy fell here, between these two houses, between my house and here."[5] That little boy was eleven-year-old Hector Pieterson, the first of hundreds injured or killed by South African police during three days of countrywide protests. The photo of eighteen-year-old Mbuyisa Makhubo carrying Hector's lifeless body as his gasping, distraught sister ran alongside splashed across newspapers around the world. For neither the first nor the last time, the UN Security Council issued a resolution condemning apartheid as "a crime against the conscience and dignity of mankind."[6] Another eighteen years of violence ensued before South Africa held its first truly democratic election in 1994.

The teenagers who rose up in June 1976 understood very well how "Bantu education" was designed to limit their abilities and ambitions. This was no spontaneous "riot." It was a well-organized insurgency, part and parcel of a long liberation struggle.[7] These teenagers had watched the foundations of systemic racism being poured, the bricks laid down day after day, not just in their schools but also in their housing, their transport, their livelihoods, and their geographies. Soweto itself had

I.1 Peter Magubane, *The Young Lions: Soweto Uprising*, June 16, 1976.
This, too, was an iconic shot of the uprising, depicting the determination of
the demonstrators in the face of the despair evident in the photo of Hector

been built as apartheid's flagship model township, largely financed by one of South Africa's biggest mining houses. The mere fact that schools for Black pupils were governed separately—indeed, the very appropriation of "Bantu" as a racial rather than a linguistic category—signaled the state's intention to colonize minds as well as bodies.

English had been bad enough; Afrikaans was untenable. As South African photographer Ernest Cole had explained to US audiences a decade earlier, "It has turned out that we studied the white man's language only to learn the terms of our servitude. Three hundred years of white supremacy in South Africa have placed us in bondage, stripped us of dignity, robbed us of self-esteem, and surrounded us with hate."[8] Cole penned those words with hope in his heart.[9] They prefaced his now-classic *House of Bondage*, the photo collection he could only publish, in 1967, by escaping from South Africa. America promised a more just society. He hadn't yet grasped how deeply his description would reverberate for Black Americans, for whom the 1933 words of historian Carter Godwin Woodson still applied: "So-called modern education . . . does others so much more good than it does the Negro, because it has been worked out in conformity to the needs of those who have enslaved and oppressed weaker peoples. . . . The philosophy and ethics resulting from our educational system have justified slavery, peonage, segregation, and lynching. . . . Negroes daily educated in the tenets of such a religion of the strong have accepted the status of the weak as divinely ordained."[10] In both South Africa and the United States, two nations built on settler colonial white supremacy, systemic and epistemic racism went hand in glove.[11]

The physical and knowledge infrastructures of "grand apartheid" distilled systemic and epistemic racism into their purest forms, weaving them into the fabric of everyday life. Cole's words resonated deeply in America: "It is an extraordinary experience to live as though life were a punishment for being black. . . . The protective institutions of society are not for you. Police, magistrates, courts—all the apparatus of the law reinforces the already absolute power of the white *baas* and his madam."[12] On these and other fronts, apartheid South Africa joined the US in epitomizing what philosopher Charles Mills calls the racial contract: the political, moral, and epistemological power relations that constitute global white supremacy.

Mills counterposes the racial contract to the social contract that Euro-American political theory identifies as a keystone of liberal democracies. Social contract theory, Mills argues, relies on abstraction at the expense of reality—especially racial reality. This is how it claims universal purview, how it sustains powerful (but illusory) aspirations to a society structured around race-less Enlightenment humanism. By contrast, racial contract theory takes full account of racism as a foundational principle of social organization in liberal democracies. This approach, Mills writes, is "necessarily more openly *material* than the social contract."[13] It refers not to an otherworldly abstraction but to political and economic reality; it's based on lived experience rather than criminally colorblind ideals.

Building on these insights, I argue that the racial contract is technopolitical. I mean this in the original and strongest sense of *technopolitics:* the purposeful design of artifacts, machines, and technological systems to enact political goals.[14] Technopolitical strategies camouflage the political dimensions of contentious issues, dissolving them into arcane technical matters that only a few experts seem qualified to adjudicate. That is certainly the case for white supremacy, which is built into infrastructures that in turn reinforce and extend racial inequality. This dynamic is particularly visible in the US and South Africa, but it holds across liberal democracies. That's why, as Charles Mills argues, "all whites are *beneficiaries* of the Contract, though some whites are not *signatories.*"[15]

Black South Africans who had to navigate grand apartheid—and the environments destroyed or remade by its infrastructures—experienced the racial contract in full technopolitical florescence. They encountered its cruelty in their work, their homes, their movements through time and space. They felt it with every breath they drew, with every step they took (or were unable to take). Even after apartheid ended officially, it remained embedded in infrastructures and environments, acquiring new life, causing new harms, and sparking new modes of resistance and refusal.[16] This book takes those dynamics as its starting point.

Some of the most powerful expressions of the racial contract in South Africa are the colossal wastes—social and sedimentary—created by its mining industry. In exploring their histories, this book pursues three broad questions. How do (these) waste histories illuminate the mechanisms of systemic and epistemic racism? How do they elucidate the infrastructural and environmental expressions of racial capitalism? And what do they teach about the nature and stakes of political struggle in the Anthropocene?

To gain leverage on these questions, I begin by developing the concept of *residual governance*. You'll read the full explanation in the next chapter, but here's a quick take. Residual governance is the deadly trifecta composed of

1   The governance of waste and discards
2   Minimalist governance that uses simplification, ignorance, and delay as core tactics
3   Governance that treats people and places as waste and wastelands

Not least because of its tight imbrication with the state—which persisted through colonialism, apartheid, and majority rule—mining in South Africa offers a prime example of residual governance. But it's far (very, very far) from the only one. Residual governance, I demonstrate, is every bit as significant and oppressive as (for example) the financial and legal instruments that have subjugated generations of Black Americans. I argue that residual governance is a primary instrument of modern racial capitalism and a major accelerant of the Anthropocene.

To make this case, I approach residual governance via those who called out its failings and sought to improve its terms, not via those who built and maintained its systems. There's no shortage of scholarship on regulation and its bureaucracies, much of which illuminates the origins of residual governance (without using the term). I've written such work myself after diving deeply into corporate and government archives. In this book, however, I mostly leave those archives—and the perspectives they reflect—behind. Instead, I ask how scientists, community leaders, activists, journalists, urban planners, artists, and others responded to the depredations of residual governance. They certainly said and did a lot. According to one activist, the documents produced about one particularly polluted catchment would, if printed and piled, exceed 5 meters in height. Stacks of such documents form the building blocks for this book. The mortar is made from my archival work and fieldwork in South Africa over the last two decades.[17]

How is it possible that despite decades of study, dozens of warnings, hundreds of studies, and major political upheavals, the residues of mining pose such a persistent problem? The question applies far beyond the case I examine. Just substitute "climate change" (or any number of other phenomena) for "the residues of mining." The story of mine waste in South Africa has many lessons for navigating planetary futures.

A straightforward chronological account is neither possible nor clarifying. Each chapter in this book follows its own temporal path, producing a narrative that loops and spirals through time. The result resembles a palimpsest, the past palpably present as new layers continue to accumulate. Chapter 1, "You Can See Apartheid from Space," offers the longest view, circling through early planetary history, human habitation, imperialism, apartheid, and the present. It also elaborates the concept of residual governance more fully.

The science, art, and activism through which South Africans confronted the residual governance of mine waste coalesced around distinct focal points: water pollution, contamination by dust, the experience of specific communities, and urban and regional planning. Each of the subsequent chapters dives into one of these themes. Chapter 2, "The Hollow Rand," focuses on drainage of acid mine wastes into the region's water sources, a common postmining problem that has swollen to titanic proportions in Johannesburg. Chapter 3, "The Inside-Out Rand," tracks responses to the even more colossal quantities of (often radioactive) dust and sand generated by mining.

The problems posed by these two types of volumetric violence—drainage and dust—were often formulated and addressed separately. But residents experienced them simultaneously. Chapter 4, "South Africa's Chernobyl?," zooms in on Kagiso township and the informal settlement at Tudor Shaft to explore how one community responded. Chapter 5, "Land Mines," pans back out to view the metropolitan region as a whole, examining how the scale and volumetric nature of residues has circumscribed the options for spatially just urban planning. The conclusion, "Living in a Future Way Ahead of Our Time," returns to the theme of planetary futures.

As you read, you'll spend time with people who live next to mine dumps and breathe toxic dust—people who survived apartheid only to live in its infrastructural detritus. You'll follow them as they claim their rights to health, home, and livelihood. You'll learn about the scientific, administrative, and bodily knowledge they required to advocate effectively for more just conditions. You'll trace the alliances among activists, community leaders, experts, and others who sought to make such knowledge actionable.

You'll see how knowledge alone, however necessary for establishing a predicament's parameters and delineating its complexity, never suffices to spur remediation or repair. For one thing, knowledge is always imperfect, insufficient, and incomplete; that's a truism, one that global industries such as tobacco and petroleum shamelessly deploy to delay regulation and deflect censure. For another, expertise is typically exclusionary. Essential questions of environmental justice take on full force here. Who counts as a knowledge producer? How can embodied, local, and social knowledge find a place in remediation plans? Finally, even if by some miracle everyone's knowledge were genuinely honored, this still wouldn't, by itself, suffice to plan adequate remediation in the face of irreversible change. Adequate for whom? For how long?

Don't get me wrong. Knowledge matters. But using it effectively also requires affective commitment, emotional engagement. Publics don't just need to know about problems; they need to feel their significance. This conviction guides my analytic and storytelling approach. My narrative wears its theory and methods lightly.[18] I strive for transparency that highlights authorial affect over academic penchants for hedging and self-positioning (penchants that I myself have indulged in past writing). After all, we live in dire times. Having experienced firsthand how scientific knowledge of harm isn't enough to generate change, environmental justice scholars have begun calling for engagement with the "traumas of living with toxic chemicals."[19] So-called dispassionate analysis is a privilege of the powerful, one that contributes mightily to our planetary predicament by allowing decision makers to dehumanize their constituents by treating them as abstractions. Keeping calm is merely a way to carry on.

Images play a major role in this story, both as illustrations and as primary sources. As an instrument of power, residual governance fosters invisibility. In response, artists, activists, scientists, and journalists have elaborated extensive visual vocabularies for calling out the injustices of this instrument. Their images add a twelfth tongue to South Africa's eleven official languages, communicating ideas and emotions that words cannot capture. Like any language, of course, images incur risks. This book deals with extreme situations; the pictures are rarely uplifting. But the point is not to invite viewers to wallow in Black suffering or indulge in ruin porn. Media scholar Cajetan Iheka writes in his masterful analysis of African ecomedia that "the image of suffering makes a claim on the viewer to act responsibly." That is precisely how the photographers I include hope their audiences will respond to their work. They seek to "*problematize* anti-Black violence," not to merely illustrate it, and certainly not to wallow in it.[20]

Engaging with these artistic vocabularies made me wonder how the arguments of this book could be rendered visually. In 2016 I participated in an Anthropocene campus in Berlin. Historian Chakanetsa Mavhunga invited me to join him and two Zimbabwean compatriots—archaeologist Shadreck Chirikure and graphic designer Chaz Maviyane-Davies—in leading a unit called "Whose Anthropocene?" I was mesmerized by Maviyane-Davies's poster series *A World of Questions*, which made compelling visual arguments concerning the ecocidal themes we were all grappling with.[21] He was kind enough to accept a commission for this book. Each chapter opens with one of his montages, presented without comment. Think of them as visual epigraphs, ways of foreshadowing an aspect of the chapter they precede.

From its inception, the mining industry formed the core of South Africa's racist infrastructures. It drove the nation's economy. It pioneered the practices that later came to constitute formal apartheid. It defined the country's relationship to the rest of the world. Massive foreign investment and expertise, especially from the US and the UK, shaped the industry from its earliest days. Starting in the late nineteenth century, American mining engineers brought the technologies and experience they'd gained extracting California's gold to South Africa's Witwatersrand plateau. The corporations they built in South Africa, writes historian Keith Breckenridge, employed more people and earned more profits than the entire US gold mining industry, until then the world's largest producer. In South Africa, American engineers found ready reception not just for their geological, mechanical, and chemical knowledge, but also for their technologies of labor organization and discipline. Their veneration of efficiency led to staggeringly racist micromanagement, down to treating African workers "like battery chickens," providing just enough "food according to scientific criteria to reach the required level of production."[22]

After the National Party rose to power in 1948, the World Bank took the lead in foreign investment. It granted $200 million worth of loans to enable the new government to expand its industrial infrastructure. The World Bank's blessing consecrated South Africa as a safe place to seek profits under apartheid. Other investors followed suit, insisting that continued industrial growth would ultimately break down racial prejudice. A central clause of the racial contract, this claim stemmed from two related delusions: (1) that racism is "a mysterious deviation from European Enlightenment humanism" rather than a constitutive component, and (2) that capitalism necessarily leads to democracy.[23]

South African Marxists coined an expression to describe the system that emerged: racial capitalism.[24] Racism, they argued, was not a side effect of capitalism. More intense capitalism couldn't eradicate racism for the simple reason that in South Africa, racism was utterly essential to capitalism. "Apartheid," wrote linguist Neville Alexander, "is simply a particular socio-political expression of [racial capitalism]."[25] Writing from exile in Britain just three months after the Soweto uprising, historians Martin Legassick and David Hemson explained that "segregation was the *means* whereby the economic interests of the mining industry were constituted as state policy."[26] They hoped their analysis would shame British firms into ameliorating conditions for African workers in their factories. Alexander, who'd spent a decade imprisoned on Robben Island with Nelson Mandela (followed by five years of house arrest), had far loftier ambitions. He imagined a future nation, Azania, which would reject the very concept of race. "'Race' as a biological entity doesn't exist," wrote Alexander, though its "social reality" was not in doubt. Mindful that rejecting race's biological reality could cut both ways, he emphasized that his vision of nonracialism was explicitly antiracist, involving not only "the denial of 'race' but also opposition to the capitalist structures for the perpetuation of which the ideology and theory of race exist."[27] In South Africa at least, capitalism was always already racial capitalism.

Such arguments resonated strongly with Black intellectuals elsewhere. Revisiting Marx's oeuvre, American political theorist Cedric Robinson concluded in 1983 that all capitalism was racial capitalism. All of it rested on racialized divisions between the free and the unfree, between valuable humans and disposable humans. As geographer Ruth Wilson Gilmore puts it, "Capitalism requires inequality, and racism enshrines it."[28] In the last two decades, she and others have reinvigorated the concept of racial capitalism, emphasizing its sedimentary structures and effects. From the seventeenth to the nineteenth centuries, these scholars argue, the kidnapping, enslavement, and murder of Africans—along with the theft of indigenous land—powered capital accumulation for white plantation owners in the Americas. Black and indigenous female bodies, writes postcolonial theorist Françoise Vergès, were "the humus of capitalism."[29] To which philosopher Achille

Mbembe adds, "Racial capitalism is the equivalent of a giant necropolis. It rests on the traffic of the dead and human bones."[30]

Starting in the late nineteenth century, North American bankers supported US military occupation and subsequent dictatorial regimes in the Caribbean, Latin America, and Asia. Historian Peter Hudson argues that they used their accumulated capital to float public debt and finance infrastructure projects, imposing "usurious rates and suffocating fiscal conditions" on supposedly sovereign nations. In the US, historian Destin Jenkins and others detail, white supremacy was baked into urban planning, real estate, and financial instruments that made it extraordinarily difficult for African Americans to build and keep generational wealth.[31] Propagated by infrastructures from plantations to prisons, racial capitalism normalized Black and indigenous dispossession in the Americas as well as in Africa. This book brings this more expansive understanding of racial capitalism to bear on contemporary South Africa.

The infrastructural violence of racial capitalism ravaged environments as well as humans. Engineer–turned–political theorist Malcom Ferdinand identifies these twinned modes of violence as a "double fracture" in planetary history, simultaneously ecological and colonial. Any serious analysis of the planet's present-day geological epoch, increasingly known as the Anthropocene, must account for systemic racism and ecocide in tandem, as processes tightly bound to each other rather than merely synchronous. Hammered in the holds of slave ships, then on plantations, then everywhere, the ecological fracture now called the Anthropocene was always also colonial and racial.[32] Black intellectuals and political leaders have long seen imperialism as "the pyromaniac of our forests and savannahs."[33] Invoking Martiniquais poet Aimé Césaire, Ferdinand argues that any environmentalism that ignores those "without whom the Earth would not be the Earth" is simply absurd.[34]

Sensitive to such arguments, some writers propose alternative appellations for the present geological epoch, designations intended to highlight the power dynamics at play in planetary change. They use terms such as *Plantationocene* or *Capitalocene* to counter the implication (which they insist is embedded in the *Anthropo-* prefix) that all humans contribute equally to the current crisis.[35] For me, however, implications of human uniformity inhere not in the term but in its deployment. Earth-systems scientists use *Anthropocene* as a capacious, multidisciplinary frame for analyzing the scale and irreversibility of human-driven planetary change. I see their term not as a declaration of war but as an invitation to dialogue, to think about the complexities of the present predicament across epistemological and disciplinary divides.[36] Why cede the term to so-called ecomodernists who proffer planetary-scale solutions like geo-engineering?[37] The word itself in no way precludes an analysis of how racism and ecocide accelerated together, feeding and shaping each other to the point that, as Vergès writes, "race became a code for designating people and landscapes that could be wasted."[38] Scholars in the emerging interdisciplinary field of discard studies have further elaborated the mutual constitution of race and waste, both as categories and as forms of violence.[39]

Long after the end of legalized racism, the proliferating residues of racial capitalism continue to sediment in financial tools, urban spaces, and health systems. Not to mention water, land, and air. These residues are rarely reducible to their molecular composition. Politics shape not only their selection, placement, and treatment, but their very chemistry; acid mine drainage is a prime example, as you'll see in chapter 2. When societies fail to design infrastructures and governance for equity and livability, then mine sulfates and their kin become the molecules of racial capitalism. By no means the only such molecules, to be sure. But significant constituents nevertheless.

Analytically and epistemologically, it makes no sense to separate racial capitalism from the Anthropocene. The residual molecules of racial capitalism drive Anthropocene accelerations. They're building blocks of Anthropocene epistemology; measuring them is what has led scientists to declare the arrival of a new geological epoch. Precisely because they materialize in infrastructures and environments, they do not require individual racists to continue their damage (though if they did, there would be no shortage of volunteers). Politically, however, policy makers and corporations have found compartmentalization extremely useful: social things like race in one bucket, physical things like molecules in another. They deploy the simplifications of modernity in order to circumscribe questions of political economy as a dichotomy between jobs and environment.

Development discourse pits employment against ecosystems the world over. Society, we are told, faces inescapable choices between jobs and environment. But there's nothing natural about this opposition. Rather, it's actively created and maintained. Neutral-sounding (dispassionate) instruments of global racial capitalism—themselves built on layers of racialized inequalities—propagate and activate the jobs/environment dichotomy. Starting with regulatory arbitrage: the corporate practice of siting polluting activities in places with weak regulatory infrastructures (weakness that itself results from the drain on human and natural resources performed by decades of rapacious colonialism and racial capitalism). In many such places, economic disparities are so extreme that jobs seem like manna, and environmental consequences beside the point. Regulatory arbitrage may not be ethical, but it's almost always legal.

By actively nurturing an alleged opposition between jobs and environment, regulatory arbitrage and related mechanisms serve as instruments of "self-devouring growth." Historian and anthropologist Julie Livingston writes that the political appeal of *GROWTH!* has turned it into a "mantra so powerful that it obscures the destruction it portends."[40] Critics who point out the physical impossibility of endless growth on a finite planet are laughed off as naive, sourpusses, party poopers who lack faith in technology's ability to fix all problems. Growth = jobs = livelihoods = justice. End of discussion.

Naturalizing a jobs/environment opposition has made it easy for many wealthy white environmentalists to imagine that poor people don't care about pollution. Or (among the more well-meaning types) that poor people have "more pressing problems" than environmental protection.[41] This may even seem like common sense. Of course people are more focused on gaining reliable access to housing, food, and basic services than on large-scale environmental threats. But what counts as large-scale? What counts as environment? Whose needs count as urgent?

Environmental justice advocates pose such questions all over the world.[42] Indeed, the movement itself emerged in response to the inadequacy of conservation-oriented environmental movements dominated by middle-class whites. Once again, Charles Mills nails it:

> Conservation cannot have the same resonance for the racially disadvantaged, since they are at the ass end of the body politic and want their space upgraded. For blacks, the "environment" is the (in part) white-created environment, where the waste products of white space are dumped and the costs of white industry externalized. Insofar as the mainstream environmentalist framing of issues rests on the raceless body of the colorless social contract, it will continue to mystify and obfuscate these racial re-

alities. "Environmentalism" for blacks has to mean not merely challenging the patterns of waste disposal, but also, in effect, their *own status* as the racialized refuse, the black trash of the white body politic.[43]

The point applies well beyond the United States. In South Africa, argues anthropologist Lesley Green, a relentless focus on species extinction long prevented many white "greenies" from recognizing Black environmental concerns as not only adjacent to their own efforts, but part of the same struggle.[44] Historian Jacob Dlamini explores the roots of this misrecognition by detailing the history of Black visitors and workers in Kruger National Park, people unseen by previous scholars who focused on Black exclusion and white conservationism.[45] In a related vein, Chakanetsa Mavhunga details the complex knowledge systems elaborated by southern Africans, arguing that their cognitive categories should be treated with the same epistemic respect as those of institutionalized science.[46] In southern Africa, as in America, simply showing that Black people have political, material, epistemic, and recreational relationships to nature is a radical move.[47]

Just as in the United States, however, Black South African environmental politics go well beyond white conservationist conventions. In 2002, scholars writing about South Africa's nascent environmental justice movement observed that access to clean water, sewer systems, and sanitation were among the most pressing environmental issues faced by Black citizens.[48] That observation still holds two decades later. Protests over lack of basic services have become more common and more intense. Some commentators dismiss service delivery protests as a minor form of politics because these focus on what appear—from the perspective of a well-plumbed bathroom—to be mundane concerns. Or because of their poop-hurling tactics, which smell far worse than your typical *toyi-toyi* (a dancing walk featured in many political demonstrations). Yet there's nothing mundane about running water for those who lack sanitation. Shit is political.

Service delivery protests are environmental (techno)politics.[49] South Africa's constitution, heralded for its progressiveness, guarantees its citizens the right to a healthy environment. Legal scholar Tracy-Lynn Humby notes, however, that the nation's constitution formulates this right separately from rights to housing and basic services, thereby entrenching old distinctions.[50] On the ground, poor South Africans (and others) don't slot their lives and struggles into environmental and socioeconomic boxes. Lived experience tells them that these are inseparable.

Residents of informal settlements, the so-called poorest of the poor, are no exception. They might not be able to detail every chemical and element making them sick. (Can you? I sure can't.) But whatever their formal education—and some have quite a lot, education alone not sufficing to prevent poverty—they possess a deeply embodied knowledge of pollution. All too often, however, that knowledge is dismissed as anecdotal evidence. Where's the method? What about the control group? Embodied experience doesn't fit neatly into the scientific protocols that underpin the process of setting regulatory limits molecule by molecule (protocols whose inadequacies are further demonstrated by the perpetual bureaucratic failure to keep up with the promiscuous production of new chemicals).[51]

What's the recourse for people without resources? Skepticism about objective scientific solutions seems warranted (and certainly easy to understand), given the long history of how racial ontologies and racist machines have shaped South African research.[52] The scale of environmental harm, coupled with huge differences in wealth and power, makes many problems seem too vast for individuals to tackle. Barring collective action, responses to individual complaints typically perform some version of the employment/environment dichotomy. "If you don't like it, just move!"[53] Yet moving is rarely a realistic option. Employees need to keep their jobs. Residents of informal settlements have nowhere else to go. Deep despair prevails, a sense that the government—in which so many invested hope in 1994—isn't on their side.

Drawing on the work of postcolonial scholars, Charles Mills argues that the racial contract both prescribes and performs epistemologies of ignorance. Consciously or not, its white adherents agree to "*mis*interpret the world," to see the world through the lens of "color blindness," which for them has the added virtue of seeming nondiscriminatory and therefore morally superior. They "learn to see the world wrongly," Mills writes, "but with the assurance that this set of mistaken perceptions will be validated by white epistemic authority." Ironically, this global "cognitive dysfunction" means that "whites will in general be unable to understand the world they themselves have made." Hence the appeal of abstract ideals and seemingly nonracial concepts, like the social contract, over material realities. White misunderstanding and self-deception, Mills insists, is not an unintended consequence; rather, it's written into the terms of the contract, "which requires a certain schedule of structured blindnesses and opacities in order to establish and maintain the white polity."[54]

Although Mills frames his analysis as a critique of white Western philosophy, its implications, as many have shown, extend much further. I find particularly strong resonance with critiques of knowledge emanating from science and technology studies (STS). That field has its own version of epistemologies of ignorance, which some call *agnotology*. This scholarship primarily engages questions of health and environment. It identifies two forms of manufactured ignorance: the malevolent kind (tobacco and fossil fuel companies actively hiding research results that are bad for business) and the systemic kind (researchers passively failing to ask questions that would, if answered, reveal extensive harm).[55] "Evasion and self-deception thus become the epistemic norm": Mills wrote those words, but they apply just as well to malevolent agnotology ("evasion") and its systemic cousin ("self-deception").[56] When these varied modes of ignorance operate simultaneously, they exacerbate environmental racism and health inequality.

Recent STS writing on ignorance urges researchers to avoid fetishizing pollution and toxicity.[57] Scholars caution that damage-centered research can exacerbate the very harms it seeks to address; focusing on toxicants without accounting for the full range of community concerns inevitably leads to false solutions. Avoiding this trap requires attending not only to molecules but also to the deeply racialized infrastructures that produce them, as well as to the ways that communities navigate these infrastructures in all aspects of their lives. One approach—which I initially intended to adopt—is to conduct research in partnership with communities. I quickly realized, however, that a great many activists and scientists in Gauteng already do this in a far more fine-grained way than I ever could. Instead, I've chosen to showcase some of their partnerships. *Residual Governance* explores their work, placing it in broader context and highlighting its intersections and dialogues with the vast body of art and journalism on mine residues.

In examining how communities and their allies have challenged residual governance, this book resists the temptations of simplification and solutionism. Instead, as Donna Haraway would say, it stays with the trouble.[58] I can only hope that my narrative enables readers to sit with the intense discomfort generated by the hard, never-ending work of repair required to survive in Anthropocenic times.

Fundamentally, the Anthropocene framework expresses trepidation about the future. What will it take for that future to be livable for humans? In these first decades of the twenty-first century, Achille Mbembe observes, many have begun to understand that "to a large extent our planet's destiny might be played out in Africa." South Africa certainly offers a dramatic demonstration of the depths and difficulty of community and planetary healing. "There is no better laboratory," writes Mbembe, "to gauge the limits of our epistemological imagination or to pose new questions about how we know what we know and what that knowledge is grounded in."[59] That's because in South Africa, the future is already here.

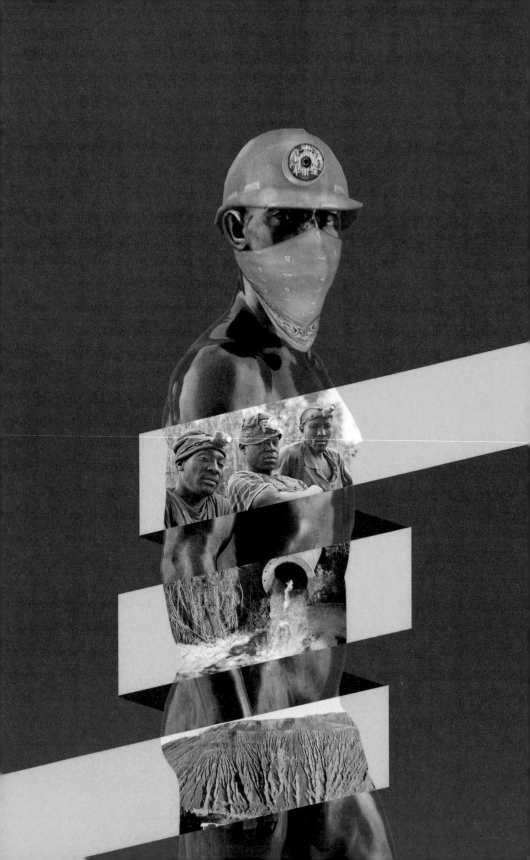

1

YOU CAN SEE APARTHEID FROM SPACE

TO FIND JOHANNESBURG and Soweto from space, look for a string of orange and yellow polygons. The largest ones tend to have all straight sides, though a few combine straight and curvy edges. Zoom in, and notice a boulevard bisecting the city, running west to east for over 8 kilometers. Continuing your descent, observe that throughout the most built-up parts of the metropolis, the polygons all sit south of this thoroughfare. Keep going until you see its name appear on your online satellite map: the Main Reef Road.

Maybe you scratch your head in puzzlement. You're looking at a landlocked conglomeration, after all. But then you start thinking on deep time scales and realize that this reef is a relic, the remains of a sea that retreated long ago: a mineral reef, not a coral reef. As you zoom

1.1 Johannesburg and surroundings, from space. Google Maps view, October 5, 2021.

in, notice the first landmark named by your online map: Gold Reef City Theme Park. Switch to 3D and continue clicking to downscale. The polygons, you now see, represent titanic tailings piles, composed of the residues of extraction. The always already fractured city was built on gold. The traces remain visible from space, consummate markers of the Anthropocene.

How big are these polygons? You decide to pay them a visit. As your plane approaches, you get a bird's-eye view of the Witwatersrand plateau, commonly known as the Rand: a nearly 100-kilometer-long band stretching west to east right through the metropolis. Over one-third of all the gold ever produced on Earth comes from this zone.[1] Its mines are the deepest and hottest on the planet, and its name was adopted as that of South Africa's currency, the rand. Your plane nears O. R. Tambo airport. In actual 3D, the piles look bigger than ever, less like abstract polygons and more like man-made mountains. An hour after landing, you've boarded the Gautrain into town. The next day, a colleague fetches you for a drive along the Main Reef Road. The stark spatial distribution of the residues acquires texture.

From above, the piles seemed lifeless. But if you turn off the Main Reef Road, you'll notice that some are inhabited. On one, you observe a group of people in white robes. At the bottom of another, nestled in a small cluster of trees, you detect some shacks. There's a lot of life along the road itself: people selling cool drinks, peddling fruit, walking home or to work. A man sells concrete bricks at one intersection. At another, a woman tends a kiln next to a substantial pile of gravel. You begin pulling over to look closer, maybe ask some questions. You may need an interpreter for some of these conversations: South Africa has eleven official languages plus a long history of migrant labor, so you never know which language you might encounter. The concrete bricks, you learn, are composed of discarded rocks and other found materials. Trash fuels the kiln, where the woman is firing up bricks made from rocks and clay.

You're still within city limits when your colleague spies an unmaintained road branching off from the thoroughfare. It leads around another pile of rocks: a hill from your current vantage point, but unremarkable from the plane. You find yourself on sloping ground. Near the top sits another brick-making operation, much larger than the ones by the side of the road but nowhere near industrial scale.

If you're foreign to South Africa, it's conceivable that until you exit the car, your experience of the environment through which you've traveled is independent of your racial identity or appearance (though not your economic class: you're in a private car, after all). Once you climb out of the car, however, that's no longer plausible.

A gaunt old white man slowly walks toward you. "Hallo," he says. *Wat gaan aan?* (What's up?) A small group of Black men, taking a break from their labor, eye you from across a courtyard. A few raise their hands in greeting. Others simply stare. Your skin tone and facial features shape their reactions. So do your clothes, your car—really, everything about you. For the rest of this story to ring true, therefore, you must imagine that you're a foreign white female academic. *Wat gaan aan?*

Your white South African colleague is an architect. In fluent Afrikaans, she explains that you're both interested in the bricks. What are they made of? Who buys them? The man takes you into his dilapidated office, pushes a cat off his desk, and describes his business. People in nearby Soweto come to him for bricks to build additions to their government-issued houses, enabling them to rent out rooms to supplement their income. Or they use them to build houses in the first place, because the government hasn't come anywhere near meeting its goal of universal housing.

The man takes you on a tour around the site, explaining the operation. Your colleague asks some pointed questions about the brick-making process. She wanders off to take photographs and collect a few gravel samples. He turns to you. *So*, he says in English, *you're from abroad?* You confirm. After a bit of small talk, he yanks his head toward your colleague, who's still prowling around. *Does she know that all the quartzite around here is radioactive?* You nod. The man is no fool. He's starting to figure out why you're here. You wonder if he asks his Soweto clients that question.

Your colleague returns, and your little group moves down the hill, which after some sloping reveals a massive, mined-out pit. Perhaps you noticed this one from the plane. A little way down the slope—not visible until you get past the rise—a group of Black men perch above a small encampment, shirts strung along a clothesline leading to makeshift shelters. The Afrikaner watches your reaction. Your colleague asks, *Zama zama?* He confirms, adding that the men in this group hail from Zimbabwe and Lesotho.

Time for another perspective. In this mental exercise, you're male; southern African, not North American or European; Black, not white.

Your life circumstances have led you to emigrate to South Africa in search of a livelihood. Hundreds of thousands have done this before you, packed onto coal-powered trains by recruiters for the world's deepest mines. Those jobs are long gone. But your brother, who used to hold one of them, knows that small bits of gold still lurk in the empty shafts. Following his advice, you assemble resources for the trip. Your brother accompanies you and your cousins. Other men bring their wives and sisters. You pay a truck driver to conceal you under a pile of goods for the border crossing, though you have a bribe ready for the guards if you get caught. After a long, bumpy, dusty trip, you finally arrive in Johannesburg.[2]

You're not welcome there. You're an alien. Locals view you with suspicion, sometimes fear. No one will hire you into a salaried job, regardless of your education and experience. Not that there are many salaried jobs to go around. But you've prepared for this. Your brother knows a cousin of a friend of an uncle who steers you to a community where others speak your language. They help you buy tools and identify a shaft that isn't already spoken for. A team of women will grind the rocks you bring up, and men will show you how to amalgamate the powder with mercury to extract the gold. You'll find a middleman to buy the output.

Finally, you're ready to descend. The work is extremely dangerous. You inhale dust and fumes. If you're lucky and find a promising vein, you may stay down there for weeks or months, in which case you'll rely on vendors to send down food at exorbitant prices. If shaft walls and roofs cave in without warning, you die. In isiZulu, *zama zama* means trying, and trying again. And again. And again. *Zama zamas* "are those who risk everything to survive."[3]

Danger also comes from other humans. So-called artisanal mining is illegal. This hasn't stopped you, your brothers, or thousands of others—but it does foster gangs, violence, and extortion. So you view the white women with suspicion. The Afrikaner tells them that the police stop by three or four times a week to collect their cut. You shift uncomfortably, glancing at your companions, then at the women's cameras. The Afrikaner notices your gaze and tells the women: *No photos.* Better. He turns back to you. *Shall we do a braai tomorrow?* You exchange jokes with him. Everyone laughs. Eventually the Afrikaner turns back to the women. *I have no problem with them*, he smiles. *We live in peace.*

At least so far, you think to yourself. You know all too well that arrangements like this can shift at the drop of a hat. For now, you're happy to share the occasional cookout. But whatever provisional trust he's earned from you doesn't extend to the white ladies. Why would it?

Violent Sediments

The tailings strung along the Main Reef Road serve as a geological index of the gold-bearing veins that once wound through the rock layers. They also function as a historical index of urban life and national development. They bear witness to the violence of colonialism, to the segregation that shaped the country's spaces and topographies. They attest to the extremely profitable racism of South Africa's mining sector, the sector that brought the nation into being and operated as a hub in global economic circuits. They are, in short, quintessential expressions of racial capitalism.

The sediments of racial capitalism frequently remain invisible to those whose lives are smoothed by infrastructures, those for whom infrastructures (mostly) work. But those smothered in the wastes of these infrastructures have no such luxury. Eternally aware of the sand in the machine, they suffer the daily effects of the grind. In present-day South Africa, this sedimentary dynamic is no mere metaphor. Especially not in Gauteng Province, which comprises Johannesburg, Soweto, Pretoria, and their surrounding areas. There, mine tailings compose the literal sediments of racial capitalism, adding new dimensions to its violence. Violence against the rock: the mining void under the metropolis is the largest on Earth. Violence against water: a century-plus of extraction has drained aquifers and, in tandem with urban development, created a perennial problem of water scarcity. Violence against hundreds of thousands of mine workers, mostly (but not all) Black, who lost lives, limbs, and health drilling through rock and hauling it to the surface. And run-

1.2 Linda Ndlovu, Daniel Mandlo, Dumisani Mahlangu, and Calvin Sibanda, 2013. These four Zimbabwean men, writes photographer Ilan Godfrey, were "highly skilled informal diggers with many years of experience," respected in their communities for their courage and skills. Godfrey visited their worksite many times to build up the trust required to photograph and name them. Godfrey, *Legacy of the Mine*.

1.3 Precious Sibanda, Roodepoort, 2013. Also from Zimbabwe, Sibanda arrived in South Africa a few months before this photo was taken. She was one of several hundred women responsible for grinding rocks excavated by informal gold diggers. Godfrey, *Legacy of the Mine*.

ning through it all, the systemic, enduring violence of racism, baked into infrastructures.

The relative placement of tailings and road were far from accidental. Starting in the late nineteenth century, mining companies, urban planners, and the state worked together to site white suburbs upwind of the mining belt, and Black townships downwind and downstream. The Main Reef Road marks this separation. Spatial arrangements built and entrenched for over a century—solidified by highways, skyscrapers, sewage lines, and other infrastructures—cannot be readily dismantled by a quarter-century of democratic elections.

That's why you can still see apartheid from space.

Tailings don't merely testify to this history; they carry its damage into the present and the future. Their most spectacular violence—tailings dam failures in 1974 and 1994 that released floods of slime with enough force to kill people and destroy homes—were construed as accidents. But their violence also operates in slower, more quotidian ways. During the winter months of July and August, winds whip across the Witwatersrand plateau, blowing toxic, radioactive dust off the piles into homes and lungs. The voids left behind by extraction also engender violence. Spectacular violence for the *zama zama* miners who descend in search of leftover bits of gold, risking entombment should an unmaintained shaft collapse on them. And slow violence, wrought by the acid drainage spilling out of abandoned mines. Water rising through the voids acidifies as it reacts with pyrite in the exposed rock face, becoming an eager host for metalloids and heavy metals (including well-known poisons like arsenic, mercury, and lead), eventually decanting onto farmland and seeping into water sources, palpably sickening people. Viewing the Anthropocene from South Africa makes it impossible to disentangle racism from ecocide.

The size of the piles, the extent of the dams, the volume of the void. The colossal scale of these residues constitutes a "wicked problem." Horst Rittel and Melvin Webber, the Berkeley professors who first formulated this notion in 1969, contrasted "wicked" with "tame." For them, "tame" problems had relatively straightforward definitions, for which experts could develop decisive solutions. Once upon a time, they wrote, the job of urban planners and architects seemed straightforward, a matter of eliminating "conditions that predominant opinion judged undesirable." For Rittel and Webber, the "spectacular" results of expert efforts spoke for themselves: "Roads now connect all places; houses shelter virtually everyone; the dread diseases are virtually gone;

1.4 Massive mine dump abutting the suburbs of Orlando West and Meadowlands in Soweto, 2011. Samantha Reinders.

clean water is piped into nearly every building; sanitary sewers carry waste from them; schools and hospitals serve virtually every district." They deemed these accomplishments "truly phenomenal, however short of some persons' aspirations they might have been."[4]

Half a century later, what first stands out is the white privilege and parochial techno-optimism embedded in their illustration of a tame problem. It's hard not to react by listing all the ways that urban spaces fell short of the ideal. How many and especially which "persons" had not seen their aspirations met? Doubtless Black Panther Party members in neighboring Oakland, who in 1970 launched the People's Free Medical Clinics along with other services to make up for white policy makers' neglect of Black residents, would have taken strong exception to the Berkeley professors' assessment.[5]

Indeed, Rittel and Webber's theory explained their own color blindness: as white urban planners, they too had a stake in the success of twentieth-century urban designs. Still, let's note that they wanted their readers to confront complexity, to question the ability of a single expert or discipline to address social challenges. They saw wicked problems as so convoluted and intractable that their very definition was contested. Addressing such problems effectively required inclusion of all stakeholders. By definition, a wicked problem could not have a single, optimal solution; what counted as a solution, optimal or otherwise, depended strongly on one's history and perspective.

The current condition of our climate presents what some scientists call a "super wicked problem." They list four features that kick problems into the "super wicked" class: "time is running out; those who cause the problem also seek to provide a solution; the central authority needed to address it is weak or non-existent; partly as a result, policy responses discount the future irrationally."[6] By these criteria, the Anthropocene is certainly super (super!) wicked. However you characterize it, one thing is clear: in the present epoch, some humans have wrought irreversible geological, ecological, and atmospheric transformations that forever change everyone's conditions for thriving, living, or—for far too many—just plain surviving. The Anthropocene marks the apotheosis of human-generated waste.[7] Especially since 1950, powerful humans and their institutions have discarded as valueless ever-increasing quantities and types of matter, matter whose release or transformation is inimical to life. Apprehending the Anthropocene means thinking about human, ecological, and geological histories simultaneously.

In many parts of the world, racialization has functioned as a permit to pollute some people's places but not others. As environmental justice scholars and activists have argued for decades, the disposability of racialized bodies forged the uneven spaces of contemporary cities, prisons, and chemical corridors. These dynamics start with extraction, which has long relied on geological science to expand its reach. Since its origins, the science of geology has been implicated in racial capitalism and ecocide.[8] This noxious entanglement continues. One recent calculation found that the wealthy countries drained $242 trillion worth of resources (including raw materials, land, energy, and labor) from the so-called global South between 1990 and 2015.[9] Another report estimates that over $40 billion of value are exported from Africa every year.[10] Such calculations put monetary values on a dynamic that activists and scholars have been highlighting for decades: Africa is not a poor continent in need of white saviors, but a rich continent whose wealth continues to be stolen.[11]

South Africa's Gauteng Province presents a particularly potent concentrate of our planetary predicament. Constituting less than 1.5 percent of South Africa's territory, the province is home to 26 percent of the country's population. That's 15 million people: 77.4 percent Black;

15.6 percent white; 6.4 percent Coloured, Indian, or Asian. Estimates suggest that 1.6 million of them live on or very near tailings dams and other mine residue areas.[12] Gauteng thus offers a dense microcosm of struggles faced by many people, in many places.

That alone would make its story worth telling. But the province offers more than a good example, more than a metonym. According to some measures, South Africa alone produces well over 80 percent of all waste generated on the African continent. In 2006, nearly 80 percent of South African waste came from mining, the vast majority of it generated on the Rand. And this is just one aspect of the story. The materials, corporations, and state entities that drive mining in South Africa have played key roles in accelerating planetary transformation and amplifying its dizzying inequities. Gauteng is both a microcosm and a motor of the Anthropocene.

Residual Governance

Wicked problems require conceptual frameworks that offer a wide variety of entry points. They demand tools for hearing and connecting dramatically different perspectives and scales, concepts that offer tools for thinking through the contradictions and multiple causalities that characterize any complex human endeavor. Such frames must themselves be wicked: to have value, they must resist easy description. The analysis in this book relies on the concept of *residual governance*, which I derive from the history, present, and imagined futures of Gauteng's mine residues. I submit, however, that the concept has relevance well beyond this case, describing many of the dynamics behind the acceleration of waste in the Anthropocene.

Consider, first, the term *residue*. In common and chemical parlance, the word refers to traces, leftovers, or by-products: the matter left behind by the main event, often—but by no means always—considered waste. Historian Soraya Boudia and her colleagues identify characteristics that unite residues of all types, including accretion (they pile up), irreversibility (you can't put things back the way they were; matter doesn't disappear but is transformed), and unruliness (even when they've been confined, residues tend to escape unauthorized).[13] The term suggests tiny particles: droplets, dust, molecules. But particle girth doesn't determine significance. Absorbing apparently minute quantities of endocrine disruptors, for example, can have life-changing consequences.

The scale of wastes can exceed that of production: burning a ton of coal produces over two tons of carbon dioxide. Tiny things can irreversibly alter ecological systems at all scales, from local springs to our planet's atmosphere.

Residues defy scalar expectations. Residue accretion is the principal physical driver of our planetary crisis. And monitoring this accretion constitutes the key method of Anthropocene epistemology: it's how we know the geological, atmospheric, and biophysical impact of human activity.

In the Anthropocene, residues are the main event.

Residual governance involves three entangled dynamics. First: governance of residues. Most straightforwardly, this involves managing discarded materials. Mining, goes the industry's inside joke, is above all a waste management project.[14] Profitable minerals typically occupy a minute proportion of their host rock, a ratio known as ore grade. The highest gold grade ever recorded in South Africa was 22 grams per ton of rock. That was in 1905. Since the late 1970s, grades haven't exceeded 10 grams per ton.[15] Concretely, that means a typical 14-karat gold chain contributes one ton of discarded rock (degraded earth) to the Rand's tailings piles—not counting the waste produced by mining the copper, palladium, and other metals that compose 40 percent of the 14-karat alloy. The residues, in other words, constitute far more material than the treasure. South Africa has recorded 6,150 abandoned mines, most of them on the Rand. Their residues continue to morph, spread, and poison. Managing these residues occupies an ever-increasing proportion of financial, administrative, and expert resources.

Second: governance as a residual activity, typically tacking between minimalism and incrementalism, using simplification, ignorance, and delay as core tactics. In a world that fetishizes commodities, the price of stuff rarely includes the costs incurred by its waste streams. New energy systems—be they eighteenth-century steam engines, twentieth-century nuclear power plants, or twenty-first-century solar panels—almost never account for the environmental impacts of extracting their fuel source or their constituent materials. Sometimes such exclusions come from the urge to simplify. Often they're deliberate. Economists call them *externalities*, a notion whose power to write off inconvenient excesses has enchanted capitalists for over a century. Originally conceived by a British welfare economist as a tool for improving social well-being by regulating negative spillover effects, the concept of externality changed valence in the hands of American economists, one of whom even received a Nobel Prize for arguing that market forces bring positive and negative externalities into "equilibrium," conveniently eliminating the need for government regulation.[16] Small wonder that capitalists fell under its spell. Words matter: the term *externality* enshrined the treatment of residues as insignificant, byproducts that required minimal attention. It legitimated ignorance by sidelining pollution-related facts and predictions.

Residues become harder to disregard when they poison bodies and land. When contamination results from a spectacular event, it gets treated as a disaster: an exceptional, one-off event that, with any luck, can be declared an act of God (thereby deflecting responsibility and potentially enabling insurance payments).[17] Exceptional events require cleanup, but don't necessarily trigger regulatory changes. Slower, systemic contamination is more insidious. Putting it on government agendas requires considerably more technopolitical work—usually by unpaid activists and underresourced communities.

Every step of the way garners fierce opposition from industrial leaders, who pay experts to help them delay regulation. Time-honored tactics include claims about "unintended consequences" and calls for "more research" (especially when their own research, kept secret under the guise of corporate confidentiality, has shown harmful effects for years or decades).[18] Once delay tactics fail, corporations collaborate with competitors on a set of best practices, then try to persuade overstretched and underpaid state experts that these best practices should serve as benchmarks for regulation.

At that point, simplification kicks in. Governance, if it takes hold at all, proceeds by reducing complex contamination pathways and contexts to a few components. This can make it impossible to apprehend negative synergies, both epistemically and technopolitically. All too often, simplification involves fetishizing linear causality: identifying a single or main cause that, if addressed, will fix the problem. This facilitates dismissing personal accounts as anecdotal, especially when they complicate simplified models. Another manifestation of minimalism: even when new regulatory regimes are enacted, their implementation and enforcement are often dramatically underresourced. And laws are worthless without enforcement. Funding growth ("growth!") is more politically palatable than funding repair. And funding episodic repair after spectacular disasters is more politically palatable than funding prolonged prevention of those disasters.[19] Residual governance in this second sense is leftover governance, that which remains after all else fails.

Third: residual governance treats people and places as waste.[20] Frontline communities and environmental justice advocates all over the world have decried this marginalization for decades.[21] From Bhopal to Durban, from Louisiana's Cancer Alley to Martinique's pesticide-infused plantations, downwind residents struggle to defend their bodies, their air, and their land against chemical invasion—a form of aggression that Portuguese artist Margarida Mendes calls molecular colonialism.[22] All too often, ordinary workers in these industries are also treated as waste dumps—especially (but not only) cleanup and maintenance crews. Consider, for example, the thousands of men who served as radiation fodder after the accidents at Fukushima and Chernobyl, some becoming so irradiated that their bodies had to be buried in lead-lined coffins. More mundanely, reactor refueling involves high-exposure work, which utilities delegate to subcontractors who don't appear in yearly employee exposure accounting, rendering invisible not only their harm but also the full exposure costs of nuclear power. Especially pertinent for our present purposes: the profitability of South Africa's vast mining system has depended on treating African bodies as waste dumps for well over a century. Since the 1880s, Black miners debilitated by their work have returned to their home villages, unable to reach mine or state hospitals. The lucky ones have families who bear the financial and emotional burden of treating silicosis, lung cancer, and mobility impairment from mining injuries: externalities one and all.[23] The less fortunate die alone.

Residual governance references all three of these dynamics and their mutual entanglement. Each has its own history. After all, the social management of residues is nothing new. Societies have always dealt with discards, with waste, with the residual legacies of their material and political pasts. But in recent times, the quantities of these residues have grown exponentially, joining the dynamics of racial capitalism with those of the Anthropocene and intensifying each in relation to the others. Problems have become super wicked, the simplifications of solutionism steadily more absurd.

Such dynamics are by no means unique to South Africa. On the contrary: I put it to you that residual governance is rapidly becoming a default mode of rule around the world.

Gauteng offers a window onto this future. Legal scholars Lani Guinier and Gerald Torres would invite us to think of the province as the proverbial canary in the coal mine, whose "distress is the first sign of a danger that threatens us all." We cannot address the challenge by "outfitting the canary with a tiny gas mask to withstand the toxic atmosphere."[24] The devil, of this and all other wicked problems, is in the details. The only way to get traction on the complexities of residual governance is intensive empirical engagement. In diving deeply into Gauteng's story, this book illuminates the complex ways in which South Africans have responded to the challenges that they (and all of us) face in a world increasingly dominated by residues. How do experts, activists, and ordinary citizens navigate the conditions of residual governance? How do they contest these conditions?

Reckoning with residual governance has become a central political task in the Anthropocene. Rather than detailing the institutions, laws, and processes through which residual governance functions (which, let's face it, makes for tedious reading), I address these primarily through the work of the scientists, communities, activists, and artists who fought against the damage caused by residues, the minimalist governance of that damage, and the treatment of people as waste. These contestations over residual governance, I propose, offer powerful tools for understanding and challenging the devilish dynamics that couple racial capitalism to the Anthropocene—not just in South Africa, and not just in the mining industry, but around the industrially entwined world.

South Africans, after all, know a lot about contestation. Their struggle against apartheid inspired activists around the world. They don't need foreigners to prescribe solutions, though they are well practiced at marshaling international allies while asserting their rights. They have been inhaling tailings dust for over a century. Rarely have they been oblivious to the ensuing damage—not since the late nineteenth century, when a botanist in charge of designing parks for Johannesburg referred to tailings piles as "poisonous mountains." In the late 1930s, novelist and poet Peter Abrahams wrote of them as pyramids "tortured and touched with the coat of death."[25] By the 2010s, media referred to Gauteng's residual landscape as "South Africa's Chernobyl."

These markers punctuate a long set of campaigns to contest the minimalism of residual governance and refuse the treatment of humans as waste. The time-honored tactic of *toyi-toyi*—a political protest dance— would not suffice. Effective contestation entailed challenging the very instruments of residual governance, including the (shoddy, selective) science used to justify its minimalism and the legal mechanisms that enacted it. This required resources. Communities needed scientists who would ask new and different questions, who would reveal critical omissions in data, who would demonstrate connections between exposure and health. They needed legal experts to identify pathways of action, file lawsuits, present at parliamentary hearings. They needed urban planners and policy makers willing and able to remake the systemic, spatial, and infrastructural instruments of residual governance. They needed media to tell their stories and keep them in the spotlight, artists who could express their pain and challenge conventional representations. And they needed activists, from within and outside their communities, to coordinate these actions and build allies at home and abroad. The minimalism of residual governance could only be countered by a flood of evidence and a relentless refusal to quit.

Communities and their allies knew all too well that secrecy was a state reflex in South Africa, both before and after 1994. They certainly weren't surprised to find that the hundreds of apartheid-era studies pertaining to toxic mine residues (and thousands of related memos and reports) remained under wraps. The new government did not make those reports automatically available. This meant activists confronted an immense informational challenge. Documents were scattered across dozens of agencies and corporations, some still staffed by the very people who'd concealed them. Fortunately, other officials had long objected to the secrecy. Slowly but surely, activists and their allies amassed archives, deploying the Promotion of Access to Information Act when needed. The recovery effort continued for two decades.[26] By the 2010s, tens of thousands of pages' worth of studies, reports, testimony, photographs, hearings, and interviews—along with countless hours of film—documented the slow but disastrous accretion of residues on the Rand. This body of evidence constitutes both the subject of this book and its primary source material.

Australopithecus and the Astrobleme

Putting human history in planetary perspective involves navigating temporal and spatial scales, often via quantum leaps rather than smooth paths. Think of this as *interscalar* travel. In science-fiction fantasies of interstellar travel, characters cover distances unbridgeable by the conventions of Newtonian mechanics, landing on worlds that teach them new ways of seeing and being. Similarly, interscalar travel on

1.5 Remining mine dumps at an industrial scale: Sibanye Stillwater, West Rand, 2022. Luka Edwards Hecht.

planet Earth elevates previously unseen perspectives and connections, sparking imagination through new juxtapositions.[27] Anthropocene stories concern geological deep time and present-day politics simultaneously. They reach back to the planet's earliest epochs, contemplate contemporary challenges, and look toward a future beyond humanity's time on Earth. They trace changes in the composition of matter itself, observing molecular transformations and navigating among subatomic particles, urban spaces, and continents. They grapple with fear, uncertainty, and anger, digging through historical detritus in search of hope or inspiration. Historians sometimes describe their work as a dialogue in the present with the past about the future. The Anthropocene demands that we dramatically extend the temporal and spatial coordinates of those dialogues.

Consider the planetary and extraplanetary forces that shaped what later became the Rand. Gold and uranium deposits began forming during the Archean eon, some three billion years ago. Sixty million years' worth of sediment into a shallow inland sea constituted the first layer. The sea retreated. More land emerged. Another 200 million years' worth of sediment accumulated. These layers might have remained inaccessible for eternity but for a colossal meteor strike. Historian Keith

Breckenridge writes of the "planet-bending energy of this collision," which left a crater some 250 kilometers in diameter: "an astrobleme, a star-wound," a blemish on the planet.[28] The meteor's impact pushed a giant mass of granite (today known as the Vredefort Dome) out of the Earth's crust, ramming gold- and uranium-bearing layers closer to the surface. That was two billion years ago. These layers weathered the supercontinental assemblages of Gondwana and Pangaea. They survived the rifts that shaped Africa and the Americas.

In time, Australopithecus roamed this place. Hominid fossils up to 3.5 million years old have been unearthed there, prompting ecstatic archaeologists to designate a zone of the astrobleme as the Cradle of Humankind, a consecration amplified by UNESCO's proclamation of the place as a World Heritage Site. You can visit this monument. Soak up the jumble of scalar claims about South Africa's significance to humanity's origins. Don't miss the gift shop's efforts to commodify deep time.

Archaeologists have found evidence for bow and arrow hunting in southern Africa that dates back over sixty thousand years, predating evidence from other areas by at least twenty thousand years. Pushing hard against stereotypes of Africans as eternally technologically backward, one team writes of the bow and arrow as a technology that requires "a high degree of cognitive flexibility: the mental ability to switch between thinking about different concepts, and to think about multiple concepts simultaneously."[29] Cognitive archaeologists use such evidence to argue that over the span of human evolution, technologies "co-create, shape, and transform our bodily experiences, brains, and genes in a continuous evolutionary interplay between our environments (environmental and social), behaviors, thinking, and biology."[30] Contemporary residues are by no means the first manufactured products to reshape human biology.

Two thousand years ago, people in southern Africa began working iron and copper. Complex polities emerged, funded by the production of metal goods. From the Shona state centered on Great Zimbabwe (thirteenth–fifteenth centuries) to the Mutapa state that succeeded it (sixteenth–nineteenth centuries) and the Ndebele state that emerged in the mid-nineteenth century (one of whose leaders, Mzilikazi, you'll meet briefly in chapter 4), these polities deployed metal goods not only for their own development but also as objects in long-distance commerce. Given the large quantity of iron goods then in circulation, archaeologist Shadreck Chirikure interprets the relatively small piles of

1.6 Australopithecus at the Cradle of Humankind, Maropeng, Gauteng, South Africa, 2015. South African Tourism, Creative Commons 2.0.

1.7 Gift shop at the Cradle of Humankind, Maropeng, Gauteng, South Africa, 2015. South African Tourism, Creative Commons 2.0.

slag left in the Great Zimbabwe ruins as evidence of a remarkably efficient reduction process that produced relatively little waste.[31]

Gold played a key role in Great Zimbabwe's Indian Ocean trade; archaeological evidence shows its extraction over a large area around the city. Some 850 kilometers to the southeast, however, what is now Gauteng offers relatively little evidence of early gold mining. Precolonial societies probably did pan alluvial deposits for gold, but heavy rainfalls would have washed away the traces. Some colonial and apartheid-era geologists (along with their fans) deemed the underground quartzite deposits "less amenable to primitive exploitation." When late nineteenth-century diggers found evidence of earlier workings, white geologists and archaeologists speculated, in their casually racist way, that these represented "traces of an ancient race, with far greater engineering skill than the Kaffirs, and civilized enough as capable of sustained and organized labour. Such relics as the diggers noticed and preserved were lost in their unsettled life."[32] (The K-word in South Africa is equivalent to the N-word in America; I will not spell it out again, but here I do so for non–South African readers to witness the epistemic violence of its casual use.) Archaeologists had proffered similar speculations about the Great Zimbabwe ruins, attributing evidence of technological complexity to the presence of "Sumerians, Phoenicians, Egyptians, Arabs, or Indians": basically, any group that wasn't Black by nineteenth- and twentieth-century standards.[33] These theories—now thoroughly disproven—fed white convictions that Black Africans couldn't possibly have created or operated sophisticated technological systems. White experts went so far as to deny the very possibility of comparing the temporal rhythms of African societies to those of Europeans. As late as 1980, one white South African archaeologist wrote that "the terms ancient workings, pre-European mines, and Iron age mines are synonymous in the Southern African context. The customary term ancient working does not imply affinities with European or Eastern cultures." Just one example among many: such statements abounded, betraying pervasive beliefs about civilizational hierarchies and human worth.[34] Today's consensus is that, following a common pattern in imperial extraction, "any trace of earlier working of these mines was probably obliterated by the first colonial prospectors, who used evidence of prior working as a convenient prospecting tool."[35] Nevertheless, erasure of precolonial history continues. In 2004, bulldozers obliterated an Iron Age site to make way for a gated golf estate "with the full complicity of the South African Heritage Association." In a 2018 survey of this history, the Bench Marks Foundation, which advocates for the rights and health of mine-adjacent communities, remarked that "like Black lives, Black history does not count for much in South Africa—both are trumped by profits."[36]

In the 1880s, white capitalists began tearing into the astrobleme's rim. Or rather, they underpaid hundreds of thousands of Black men—migrant workers from across southern and central Africa—to do so under the supervision of a few thousand whites. To manage recruitment, the Chamber of Mines created the Witwatersrand Native Labor Association: WNLA in white documents, Wenela among Black workers. Historian Jacob Dlamini tells how Wenela runners working Portuguese territory (in present-day Mozambique) escorted recruits to processing camps in Kruger National Park, where they would receive a passbook and passage to the mines. Some prospective recruits found ways to work around this system. In 1948, for example, one official complained that "natives" were pouring in "without restraint," having "found avenues of employment where wages are as high or higher than on the mines." Avoiding Wenela meant their earnings would remain intact, not subject to "deductions on account of fares, customs, etc."[37] Dlamini describes how these "clandestine emigrants" shared information to facilitate the journey, as would their twenty-first-century descendants in their quest for *zama zama* work.

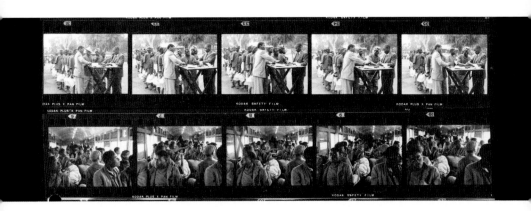

From the processing camps, migrants boarded trains to the gold, diamond, and coal mines where they would spend the next six to nine months. Jazz musician Hugh Masekela recounts the infamous journey in his iconic 1971 composition "Stimela," which became an anthem of the anti-apartheid movement. "There's a train that comes from Namibia and Malawi," he begins. "From Zambia, Zimbabwe, Angola, Mozambique, Lesotho, Botswana, Swaziland . . . young and old," he intones, these men work the mines "sixteen hours or more a day for almost no pay." His voice drops further as he follows them down, "deep, deep, deep down in the belly of the earth . . . digging and drilling that shiny mighty evasive stone." When they are served "that mish mash mush food into their iron plates with the iron shank, or when they sit in their stinking, funky, filthy flea-ridden barracks," men think about their families, "their lands, and their herds that were taken away from them with the gun, and the bomb, and the tear gas, the gatling and the cannon." If you've never heard the song, I urge you to play one of the dozens of available versions to hear Masekela's rage and rhythm, his howling humor and haunting trumpet.

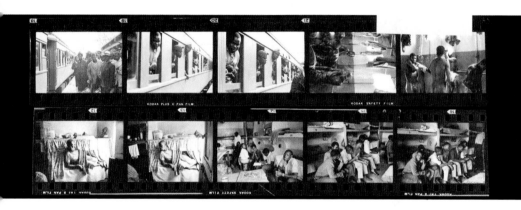

Mining requires hard labor the world over. Nevertheless, few workplaces compared to the Rand's hot, cramped stopes (as South Africans called the narrow shafts). When rocks burst and shafts collapsed, the lucky workers were those who lost only a limb. Miners inhaled silica dust and radon, with terrible consequences for their lungs and lives. Cyanide, used to leach gold from rock in aboveground processing plants, seeped into the water table. But the digging continued, relentlessly, powered by men who endured daily humiliation underground and above, using hand tools long after mechanization prevailed else-

1.8-1.9 Recruitment, travel, and arrival at the mines. Contact sheets for Ernest Cole, *House of Bondage*, 1967. Hugh Masekela and Cole were among the most prominent Black South African artists to grapple with traumas of mine work. See chapter 3 for a fuller discussion of Cole and his work. Ernest Cole/Magnum Photos.

where. On the Rand, human energy was cheaper, especially when the humans providing it could be treated as disposable. Mechanization did, however, serve to pump water out of the shafts as they plunged below the water table: even expendable men couldn't dig underwater. The mines rapidly became the deepest in the world. By the 1930s, hundreds of tunnels penetrated the rock basin, connecting underground in three main groupings: the West Rand, the Central Rand, and the East Rand. Aboveground, the mountains of discarded rock grew, and grew, and grew.

Labor practices developed by the mining industry in the first half of the twentieth century provided prototypes for racist legislation in the second half. Apartheid's infamous passbooks (known unaffectionately to their bearers as the *dompas*, or "stupid pass") began as recruitment and surveillance tools on the mines. Companies housed workers in compounds, where rooms featured concrete bunks stacked four high, sanitation facilities were minimal, and food consisted of slop (Masekela's "mish mash mush"). The mining sector didn't just set a template for apartheid laws and practices. It also funded their implementation. In 1956, for example, mine magnate Ernest Oppenheimer contributed £3 million to help the state clear Black residents from "white areas." Half a dozen mining houses followed his example, eager to curry favor with apartheid politicians.[38]

Outside the compounds, colossal piles of waste rock demarcated urban space, serving as buffers that would separate settlers from Africans. City planners worked tirelessly to relocate (read: forcibly remove) Black residents to townships south of the mine belt. Immigrants of Chinese and Indian descent, along with mixed-race (Coloured) residents, barely fared better. South wasn't just downwind of the piles. It was also downhill. Soweto sat in a basin below the Central Rand mines. Watershed runoff coursed through mines and their discards, down into the township. For planners, this flow made Soweto's Klipspruit district an ideal location for Johannesburg's sewage works.[39]

White settlers imagined such exclusions as eternal. Cemeteries cemented discrimination in death, separating sepulchers first by race, then by faith. On one side, the remains of nonwhite bodies, divided into racial groups labeled by white racist nomenclature, with "Christian K-s" getting a separate plot from other "K-s." On the other, the remains of white bodies, divided into mostly Protestant sects. Down the middle, buffering white Christians from everyone else: Jews, a swath of society that (as writer Mark Gevisser later remarked) ran the gamut from "mining magnates to pimps."[40]

Infrastructures inscribed and imposed separatist ideologies. Settlers forced Africans to live with the wastes of whiteness, breathing its dust and drinking its effluents. For white settler society, Black people were waste, despite their value as cheap labor.

Tailings, then, did not merely testify to the extraction of elements. They were also active agents of colonialism and its apartheid accomplice. In today's South Africa, their governance presents a familiar problem of boom-and-bust capitalism: many of the companies that produced this waste have shut down, leaving only a few accountable. But their residual governance also reveals the deeper dilemmas that accompany profound political change, here produced by the country's first democratic elections in 1994. What to do with discards after the end of a nation? To what extent, and how, should the "democratic dispensation" pay for damages wrought by its predecessor?

The difficulty, of course, is that histories don't end. They accrete. Apartheid may no longer exist as a government-run project of oppression. But its practices and effects remain embedded in South African life. Nearly three decades into the new nation, political scientist Sipho Mpofu-Walsh is one of many who find severe deficiencies in the democratic dispensation. Systemic exclusions enshrined in spatial, legal, financial, technical, and carceral assemblages, he argues, operate as an "algorithm running in the background of South African society."[41] Scholars may disagree about whether such continuities represent a "new apartheid." But there's no question that apartheid's physical infrastructures continue to shape contemporary life in ways both deliberate and unplanned, much like an algorithm. Despite the emergence of a new Black elite, the nation's fundamental racial contract has changed very little. Apartheid's technopolitical forms enable it to persist in zombie form.

Similarly, the material relics of our Anthropocene present will endure long after nation-states have disappeared. So it's particularly pertinent to examine a place where one nation-state ended and another one arose—one which, at least for a while, truly tried to offer its population better lives. South Africa offers a glimpse of life after the end of a world.

A Curie of Rock

By the late twentieth century, 50,000 tons of gold had been ripped from the ground. That's about a third of all the gold ever mined on planet Earth. Accumulated over 260 million years. Dormant for two billion more. Gone in one century.

Poured into ingots, South African gold powered capitalism around the globe. It amassed in bank vaults. It buttressed the gold standard.[42] It financed racialized economies.[43] Fashioned into Krugerrands, it shone as a public relations tool for the apartheid state, insinuating itself into imaginations: in 1980 alone, US citizens gobbled up nearly a billion dollars' worth. Political scientist Willard Johnson (MIT's first Black professor) noted that this represented "more than the entire US trade deficit with South Africa for that year." Johnson and other African American activists made Krugerrand protests a centerpiece of their anti-apartheid campaign.[44]

Uranium also lurked in the astrobleme's layers, trapped in the same ore matrix as gold.

The clearest description of the relationship between uranium and gold, and the resulting radioactivity of rocks, comes from G. Wendel, a mine expert who worked on the West Rand. He invites readers to imagine a block of Rand rock 100 meters long, 100 meters wide, and 1 meter thick. The mass of this rock would be about 27,000 tons, of which 3.2 tons would be uranium, 135 kilograms (or 0.135 tons) would be gold, and 1 gram (0.000001 tons) would be radium. The first unit used to measure radioactivity was named the curie, after the celebrated French scientists, and defined as the radiation produced by 1 gram of radium. Accordingly, Wendel invites us to think of this hypothetical block as a "Curie of Rock."

Uranium is radioactive. This means that its atoms aren't stable: they decay into other isotopes. These "progeny" also decay. Along the way, atoms emit alpha particles (consisting of two protons and two neutrons), beta particles (fast electrons or positrons), or gamma rays (high-energy photons). One progeny is particularly notorious: radon. Readers may know it as a home hazard, a gas that lurks in the basements of buildings built on bedrock containing even trace quantities of uranium. As an alpha emitter, radon is particularly pernicious. The gas and its progeny enter your lungs, where they continue to transform. The resulting radiation damages your DNA. Radon is the world's second leading cause of lung cancer after cigarettes.

I'll have more to say about radon exposure as I go along. For now, let's get back to our block of rock. Ultimately, uranium's decay chain ends in a stable isotope of lead. "Every second, in this Curie of Rock," Wendel explains, "3.7×10^{10} [37 billion] atoms of uranium and each of its daughter products decay by emitting either an alpha or a beta particle." A radioactive element's half-life equals the amount of time it takes for half of its atoms to decay. Half-lives range from nearly 4.5 billion years (for uranium 238, so-called natural uranium) to 16 milliseconds (for its decay progeny polonium 234). This means the rock's composition is changing constantly. On a human timescale, the rock appears stable; short-lived progenies disappear fast, but with more constantly being generated higher up in the decay chain, you get an equilibrium of sorts. On a geological timescale, however, it's a different story. After 4.5 billion years, Wendel concludes, "half of the uranium atoms present in the Curie of Rock will decay and be converted into lead. The current Curie of Rock will have turned into a half Curie of Rock."[45]

As long as the rock remains undisturbed, its radioactive elements have little opportunity to wreak mischief. But "as mining of our Curie of Rock starts, there is initial fracturing of the rock, which allows air to penetrate the rock, and allows radon gas which was trapped in the rock to escape into the air-filled cracks in the rock." This poses a serious problem if you're an underground miner. It also poses problems when you pile broken, uranium-rich rock into colossal mountains of tailings. Which is how the Rand's uranium was handled during the first six decades of industrial-scale gold extraction.

Nuclear weapons changed the fate of Rand uranium. After their first use in 1945, the US and the UK lost no time signing uranium procurement contracts. Continuing their long association with South African mines, US metallurgists helped to design the country's first uranium ore–processing plant. Built to separate uranium from other mine tailings, the plant was inaugurated in 1952 with great fanfare. Sixteen more plants quickly followed. The first 10,000 tons of South African uranium oxide originated in the gold tailings of the six preceding decades.[46]

Had you followed early uranium shipments out of apartheid South Africa, you would have landed in the US, the UK, France, or Israel. The chase would have led you to enrichment facilities, weapons manufacturing plants, and missile silos. Some of this uranium fissioned in nuclear weapons tests, transforming its atoms into other radioactive isotopes. These isotopes infused oceans, wafted across continents, settled on sands. Some lurked in the atmosphere, later providing scientists with evidence of our warming planet.[47] Indeed, one contender for the Anthropocene's "golden spike"—the signal marking transition from one geological epoch to another—is July 16, 1945, at 5:29 a.m., when the US detonated its first atomic bomb in New Mexico. Stratigraphers identify this explosion—together with its successors in Japan, the Marshall Islands, the Kazakh steppes, the Nevadan and Saharan deserts, French-ruled Polynesia, and the lands of the Maralinga Tjarutja people—as signals of the Anthropocene's golden spike because their collective radioactive signature appears everywhere on the globe.[48]

Not all South African uranium ended up in nuclear weapons. Some went to nuclear power stations, where spent fuel continues to accrue pending a nuclear waste disposal solution. Some collected in the waste streams of enrichment plants. Revalorized as depleted uranium, yet still radioactive, it became ammunition for antitank penetrators. Ever the nuclear pioneer, the US first deployed these weapons in the 1991 Gulf War. They've since appeared in Kosovo, Iraq, and Syria, where their use has been linked to birth defects and other illnesses.

Back in Gauteng, uranium and its progeny continued to permeate bodies and land. This infiltration did not make news until the early 2000s. The most arresting headline, from 2011, topped a story about informal settlements adjacent to tailings piles: "Living in SA's Own Chernobyl."[49]

The paper's readers could well conclude that this was new news. But industry insiders cannot be so easily forgiven. The health dangers posed by uranium-bearing rocks were first identified by Agricola, the sixteenth-century mineralogist often considered the patron saint of mining. In 1975, three-quarters of homes in the US town of Grand Junction, Colorado, were declared radioactive and in need of demolition: built with uraniferous tailings, their walls emanated radon. During the decades of formal apartheid, South Africa's Atomic Energy Board declared time and again that radon in mines was innocuous, actively dismissing evidence from elsewhere and repeatedly refusing to collect its own.[50]

Industrialists love to pretend innocent ignorance to explain environmental harm. "We couldn't have known back then," they say, insisting that revelations of harm flow from recent (ideally "uncertain" or "incomplete") science. This classic move isn't any truer for radio-active contamination than for the carcinogenic properties of tobacco or the climate-warming properties of fossil fuel combustion. Or, as you're about to see, for the water-acidifying properties of abandoned mine shafts.[51]

2

THE HOLLOW RAND

JOHANNESBURG IS THE ONLY megalopolis in the world that did not spring up around a major water source: so begin many policy briefs about Gauteng's water supply. This rhetorical gesture renders the province's present-day hydration challenges as the unintended consequence of growth in an already arid land. But the story is more complex. The Witwatersrand constitutes southern Africa's largest watershed. Watercourses to the east of the ridge flow into the Indian Ocean, while those to the west flow into the Atlantic. Communities and subregions bear traces of water in Afrikaans toponyms ending with *-fontein* (source) or *-spruit* (spring), small waterways by some measure, but vital to the communities and ecologies they serve. Their vast spread hints at the sprawling aquifer that subtends Gauteng and neighboring provinces.[1] As recently as 1937, one commentator could still describe the ridge as "well-watered."[2] It was the acceleration of Anthropocenic transformation that created scarcity.

That acceleration required transforming water into an object of governance. This was the focus of the region's first formal governing body: the Rand Water Board, created in 1902, two years before the incorporation of Johannesburg as a municipality. Governance and planning structures proliferated atop and alongside this one throughout the twentieth century, buttressed by a legion of consultants peddling their hydrological, mineralogical, and pedological expertise. From the beginning, governance on the Rand entangled private expertise and public policy.

The Rand's water vigorously resisted infrastructural manipulation. When mine shafts descended into the aquifer, water rushed into the cavities. Operating the mines required pumping out the liquid. At peak mining, companies sucked out some 20,000 cubic meters of water a day from the Rand's Western Basin alone. Farmers in the region had long protested this practice of "dewatering," which affected crop irrigation and polluted the region's streams with effluents.[3] Growing mine voids engendered catastrophic sinkholes, which could gobble up animals, crops, and (on occasion) people. Farmers also resented the mines' aboveground behavior, which sucked up water to feed the treatment plants that leached gold from rock with cyanide.

One potential source of water for the Rand lay in the highlands of Lesotho, a tiny landlocked country perched atop mountains south of Johannesburg. Generations of Basotho men had migrated to work on the Rand mines, where they leveraged apartheid's ethnic stereotypes to cultivate a reputation as uniquely skilled shaft-sinkers who merited higher pay.[4] In 1966, the year that Lesotho won formal independence, nearly 10 percent of the country's population worked on the mines. Eyeing the potential for Lesotho's apparently abundant water to supply the industries and inhabitants of the Rand, the apartheid state backed a military coup in 1986. This paved the way for the Lesotho Highlands Development Project. Political turmoil in 1998, the first year of water exports, prompted the African National Congress (ANC) government to send troops to secure one of the project's main dams with deadly force, demonstrating beyond all doubt its commitment to the apartheid-era arrangement.

Sunny rhetoric about Lesotho's national independence notwithstanding, the people resettled by the project experience it as a conflation of South African neocolonialism with World Bank neoliberalism. Anthropologist Kefiloe Sello, herself among the displaced, documents the price paid by ordinary Basotho. The commodification of water compromises their ability to live, let alone thrive, in relation to their environments: not only have they lost vital medicinal plants to the scrape of the excavators, but many have less access to water, not more. Lesotho water hasn't made a tangible difference for the residents of Gauteng's informal settlements either. In both places, people queue at communal taps, sometimes for hours, to collect water in buckets.[5]

Before and after apartheid, the manufactured hydrological system that bound Lesotho and Gauteng served mining and other powerful interests above all else. This became especially evident when Rand mines and their pumping stations began to shut down. Hundreds of kilometers of empty tunnels pierced the now hollow Rand, lying in wait for the return of water, ready to change its chemical composition.

The magnitude of these mine voids is difficult to grasp. Dozens of mining companies had drilled through rock to follow the gold seams. The plateau hosted six of the world's deepest mines, one of which plunged over 4 kilometers underground.[6] Although excavated in dozens of separate operations, the shafts connect underground, where they coalesce into intricate warrens, one in each of the three basins. The volume of the West Rand void alone is around 45 million cubic meters. Unlike the tunnels of Gaza and the West Bank described in architect Eyal Weizman's *Hollow Land*, the shafts of the hollow Rand were not explicitly imagined as weapons of occupation by their designers.[7] But that's how they have been experienced by generations of workers and residents. In Palestine, the hollow land enabled frequent, spectacular violence. In Gauteng, the hollow Rand channeled slow violence.

The hollow Rand turned water into a stealth weapon. When a shaft shut down, its operators stopped pumping. This left remaining companies to shoulder the burden of draining the entire void. As the pace of shutdowns picked up, these operators lacked resources to handle the load. For a while, the state offered pumping subsidies to help them stay profitable. Eventually, however, the pumping stopped. Slowly but implacably, water rose through the stopes, oxidizing pyrite in the exposed rock face. The ensuing chemical reactions acidified the water, making it an eager host for an abundance of metalloids and heavy metals, including well-known poisons like arsenic, mercury, and lead.

Some experts had seen this coming. For decades, they'd warned about the "decant": acid water gushing onto the surface. But in the absence of visible discharge, their alarms went unheeded, relegated to a distant future by the minimalism of residual governance. As one group of ex-

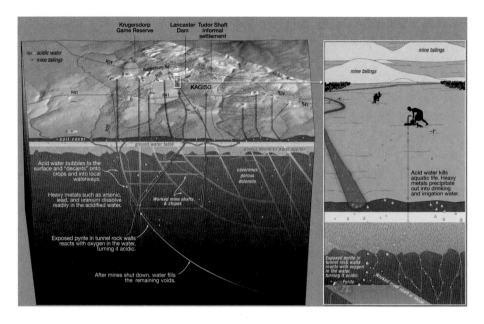

Labels within figure (left panel):
acidic water
mine tailings

Krugersdorp Game Reserve
Lancaster Dam
Tudor Shaft informal settlement

R24
Rustenburg rd
R28
KAGISO
R41
R41
R24
R41
R28

soil cover
ground water table
porous dolomitic karst aquifer

Acid water bubbles to the surface and "decants" onto crops and into local waterways.

cavernous porous dolomite

Heavy metals such as arsenic, lead, and uranium dissolve readily in the acidified water.

Worked mine shafts & stopes

Exposed pyrite in tunnel rock walls reacts with oxygen in the water, turning it acidic.

After mines shut down, water fills the remaining voids.

Labels within figure (right panel):
mine tailings
mine tailings

Acid water kills aquatic life. Heavy metals precipitate out into drinking and irrigation water.

Exposed pyrite in tunnel rock walls reacts with oxygen in the water, turning it acidic.
Pyrite
Worked mine shaft or stope

2.1 Acid mine drainage in the West Rand. Designed by Paula Robbins.

perts noted ruefully, the problem was "ignored, rather than unforeseen."[8] Not until 2002, when the noxious concoction decanted from the shafts—flooding land, contaminating food and water, sickening people and animals—did acid mine drainage (AMD) receive official recognition as a residue in urgent need of governance.[9]

Which, in turn, presented several proximate problems. Not the least of these involved the familiar conundrum of boom-and-bust capitalism: many of the mining companies (or at least their operational subsidiaries) had shut down, leaving few to be held accountable. Endlessly repeated the world over, this take-the-mineral-and-run move has been a major motor of Anthropocenic dynamics. The move took on heightened salience in South Africa, where it wasn't just the companies that had done a runner, but the state itself. The new democratic government inherited the toxic residues and volumetric violence of racial capitalism, along with the management mechanisms that had characterized residual governance under apartheid.

An Old Problem

The sheer volume of acid water gushing out of the hollow Rand was unprecedented. But the dynamics and consequences of acidification had been observed for millennia. In 3000 BCE, a bright red river in to-day's southwest Spain caught the attention of Tartessians and Iberians, who began scratching around its banks for minerals. Later known as the Rio Tinto (red river) because of its hue, the watercourse continued to attract attention. Phoenicians named it Urbero, or "river of fire." Then came Greeks, Romans, Visigoths, and Moors. Spaniards mined its banks in the eighteenth century, followed by British companies in the nineteenth; their success led to the formation of the eponymous Rio Tinto mining conglomerate.[10] Scientists now hypothesize that microorganisms capable of oxidizing sulfur and iron created the extreme conditions betrayed by the river's color. But no one doubts that the river's long history of mining also contributes to its present pH of 2. (For those not chemically inclined: the fourteen-point pH scale indicates acidity and alkalinity, where 7 is neutral; 2 is ultra-acidic.)

Southwestern Spain was by no means the only place with long-standing, visible acid mine drainage. By 1550, when German physician and mineralogist Georgius Agricola finished his legendary treatise on mines and minerals, the deleterious effects of mining were well known. So much so that Agricola chose to begin *De re metallica* by enumerating contemporary objections to the entire enterprise. Mining, said sixteenth-century critics, relied on "sordid toil." Its products benefited only a small number of people, "because forsooth, gems, metals, and other mineral products are worthless in themselves." Critics lamented the dig's deadly outcomes: "Miners are sometimes killed by the pestilential air which they breathe; sometimes their lungs rot away; sometimes the men perish by being crushed in masses of rock; sometimes, falling from the ladders into the shafts, they break their arms, legs, or necks." But Agricola insisted that "things like this rarely happen, and only in so far as workmen are careless."[11] Blaming operators for systemically induced disasters has a very long history.

Agricola did concede that mining could devastate fertile farmland and forests. The enterprise exacted "an endless amount of wood for timbers, machines, and the smelting of metals. And when the woods and groves are felled, then are exterminated the beasts and birds, very many of which furnish a pleasant and agreeable food for man." Water used to wash the ore "poisons the brooks and streams, and either destroys the fish or drives them away." Such devastation (his word) made it tremendously difficult for inhabitants to procure "the necessaries of life."[12] But none of this meant that mining produced more harm than value. As you read the following passage, consider that it was translated from the Latin in 1912 by none other than mining engineer (and not-yet US president) Herbert Hoover and his wife Lou:

> If there were no metals, men would pass a horrible and wretched existence in the midst of wild beasts; they would return to the acorns and fruits and berries of the forest. They would feed upon the herbs and roots which they plucked up with their nails. They would dig out caves in which to lie down at night, and by day they would rove in the woods and plains at random like beasts, and inasmuch as this condition is utterly unworthy of humanity, with its splendid and glorious natural endowment, will anyone be so foolish or obstinate as not to allow that metals are necessary for food and clothing and that they tend to preserve life?[13]

Deeming metals—and therefore mining—essential to civilization regardless of consequences, Agricola insisted there could be no turning back.

No wonder that the mining engineer who would become president, then conservative icon, was enchanted. Agricola's paean to civilizational progress led directly to the gospel of efficiency that Hoover had preached as a labor consultant on the Rand a few years earlier. Centuries of history justified sacrificing the environment in the name of progress and growth.

And also: no self-respecting mining engineer in twentieth-century South Africa—or anywhere else—could legitimately argue that acid mine drainage was an unanticipated consequence of mineral extraction.[14]

Acidified Apartheid

Black South Africans bore the brunt of apartheid's violence. The slow violence of environmental contamination also affected whites, however, especially those at the economic margins. This predated apartheid, of course. But the regime's growing obsession with secrecy—which only intensified with the imposition of international sanctions and South Africa's nuclear weapons development—cultivated bureaucratic practices that minimized public communication, leaving even white citizens in the dark about threats to their health and livelihood.[15]

The expansion of water and electricity infrastructures across the Rand in the first half of the twentieth century had enabled mining houses to adopt new milling techniques. Instead of sand, mills now produced fine-grained composite slimes (an industry term), the residues of which were pumped into "slimes dams." The finely ground slimes released more sulfur from the pyrites than the sand dumps did, making their residues even more acidic. By 1940, the industrial advantages of slimes had led many mining operations to abandon most sand dumps in favor of slimes dams.[16]

Fueled by coal-burning power plants, the postwar boom greatly accelerated pollution of the region's water. In 1948, the Water Board's chief engineer expressed "grave concern at the quantities of obnoxious effluents entering the Vaal barrage . . . the most objectionable feature being the addition of sulphates."[17] Sulfates acidified water, making it more hospitable to heavy metals. That same year, white farmers on the West Rand lodged a complaint about effluents in their water sources. A survey of the Vaal River catchment begun in 1952 confirmed extensive pollution. In 1956, a dam burst at the Grootvlei mine, east of Johannesburg, released one-third of the impoundment's contents. The following year, the Chamber of Mines itself issued dire predictions of deteriorating water quality in the absence of mitigating measures. By the end of the 1950s, springs had dried up in some areas, and farmers were forced to use mine discharge water to irrigate their wheat crops and water their animals.[18]

These growing problems stimulated the passage of the Water Act in 1956. This was the first piece of legislation that required water users to minimize—and pay for—water pollution. But legislation is one thing, enforcement another. Dam bursts were easy to detect. Fully tracking the continuous release of pollutants, however, demanded a dense monitoring network. How would pollution be demonstrated? What counted as evidence?

In 1960, Gold Fields, one of the larger mining houses, commissioned the national Council on Scientific and Industrial Research (CSIR) to study water flow into sinkholes and irrigation channels. The grim results of this investigation remained confidential, accessible only to Gold Fields and other mining houses.[19] Four years later, the director of the National Institute for Water Research finally laid eyes on the confidential reports. He immediately called for water quality control and pollution prevention planning, but his appeal fell on deaf ears.

Farmers didn't need a report to know that their wheat crops were failing, or that their animals suffered miscarriages, birth defects, internal bleeding, milk coagulation, and, in the case of at least one pig, bone cancer. The water supply for the Blyvooruitzicht mining village, home to South Africa's first uranium extraction plant, was so awful that even coffee made with it tasted dreadful. Several witnesses reported that disproportionate numbers of Blyvooruitzicht children had "mental impairment."[20]

In late 1967, the region's residents reported their troubles to the deputy minister of Agriculture and Water Affairs. A veterinarian who examined the animals reported that they suffered from extreme calcium deficiencies, probably caused by exposure to heavy metals and radioactive isotopes. The mining houses countered that in the absence of tests revealing radioactivity in mine water, it was "dangerous and irresponsible" to blame the problems on radiation. Siding with the mines, one Water Affairs official accused the farmers of lying about changes in water sources and flows—an accusation he was forced to retract after one of his colleagues confirmed the farmers' statement in a separate investigation. The ministry tasked a committee, chaired by a CSIR scientist and including representatives from the Chamber of Mines and the Atomic Energy Board (AEB), to investigate farmers' concerns.

Based on a grand total of three water samples, the Chamber of Mines reported that water radiation levels did not exceed background. An AEB study, however, yielded levels of alpha and beta radiation that exceeded permissible limits. J. K. Basson, the AEB representative, hastened to reassure the industry, falsely claiming that international consensus agreed that existing permissible limits were ten to twenty-five times too conservative. He disdained anecdotal reports as fearful and ignorant, and insisted there was no need for further investigation.[21] Basson would go on to an illustrious career as an ignorance peddler, repeatedly dismissing the danger of radiation and denying the need to study its manifestation in mines.[22]

The region's mine operators rushed to purchase the affected farms. One geologist who lived in the area noted that this "effectively silenced" the community; the "sole remaining voice was considered . . . a trouble maker." By 1971, when the committee presented its final report to the deputy minister, "there was no community left to accept or reject its conclusions."[23] The problem fizzled out as a political issue, and uranium- and radium-bearing water continued to flow into groundwater and sinkholes.

Another two decades would elapse before the few remaining farmers renewed their complaints about water quality to Gold Fields. Denying responsibility, the company warned farmers that it had "deep pockets" and could easily keep them in court for sixty years. Undeterred, the farmers again appealed to the Department of Water Affairs, which—despite stonewalling by one decidedly unsympathetic official—agreed to investigate. The inspection confirmed problems with water quality. The mining industry launched its own investigation, only to learn that more sophisticated techniques and larger sample sizes corroborated the finding: few samples registered levels of radium 226 below the prescribed limit of 5 picocuries/liter. "It is possible," they concluded, "that many mines will require an extensive monitoring system for some distance downstream of the discharge point." Predicting negative publicity, investigators recommended treating discharge water to bring contaminants down to "an acceptable level," but added that this depended on finding a "cheap, simple treatment technology." Meanwhile, they advised treating "the outcome of this work . . . as highly confidential." Rather than releasing the results to the public, these "should be used to forewarn the industry . . . and to assess the technical and economical [*sic*] feasibility of minimizing the impact of [water-related exposure] pathways."[24]

South African mining companies thus joined the legion of industrial powerhouses—most notoriously in the tobacco, asbestos, and fossil fuel sectors—that buried scientific findings about health and environmental harm.[25] As in so many other instances, words like "feasibility" and "acceptability" recast the unsavory pursuit of profit into neutral technical terminology. (Dispassionate!) Bottom line: the industry would implement only the most minimal mitigation measures. The state had neither the capacity or (for the most part) the desire to insist otherwise. In short, textbook residual governance.

Democratic Dispensation

South Africa's transition to democracy in 1994 had mixed effects on its mining industry. The lifting of sanctions certainly eased the sector's technopolitical burdens. But a truly democratic dispensation threatened the profits enabled by residual governance by (potentially at least) dramatically intensifying regulatory oversight. There was certainly plenty to police. Within a few years, a host of new regulations emerged, covering water, mining, waste, land, and radiation. These, in turn, were administered by a dense web of agencies and departments. Standard setting for water quality occurred at the national level; municipal agencies were responsible for water and sanitation services; and provincial authorities covered a range of issues in between.[26] A full description of these intricacies could easily take up an entire (exceedingly boring) book. For now, it's enough to gesture at the complexity of the bureaucratic terrain citizens had to navigate in order to assert their constitutional right to a clean environment.

Overcoming residual governance promised to be difficult and complicated, not least because of the complex cocktail of contaminants that flowed through Rand water. These included well-known toxins like arsenic, mercury, and lead. But as democracy dawned on South Africa, the metal that received the most attention was uranium. And nowhere was it more concentrated than in the West Rand, the main source of South Africa's uranium exports during the apartheid period. There, uranium-laced water flowed freely into the Wonderfonteinspruit catchment. (Wonder-fontein-spruit: wondrous-source-spring.)

The regulatory threat to mine companies manifested almost immediately, in the form of a revamped Nuclear Energy Act that gave nuclear authorities oversight of radiation protection in mines.[27] The industry mounted a vigorous challenge, arguing that radiation exposure was "essentially a health issue and not a nuclear energy issue" and should therefore fall under the purview of the Department of Health (which harbored no nuclear expertise).[28] In a 1995 petition to Parliament, Chamber of Mines president A. H. Munro invoked the International Commission on Radiological Protection (ICRP) recommendation to keep radiation exposure "as low as reasonably achievable." He explained that ICRP dose limits included "social judgements . . . [that] would not necessarily be the same in all contexts and, in particular, might be different in different societies."[29] These euphemisms elided a tacit assumption about what historian Michelle Murphy calls "the economization of life": namely, that human lives have different monetary value in different places.[30] Vigorously denied by the experts who wrote exposure recommendations, this assumption nevertheless underpinned phrases like "reasonably achievable" and "social judgement." Reasonable for whom? Whose judgment had the authority to stand in for all of "society"?

The Chamber insisted that nuclear regulation of mines would impede economic and social development in the new South Africa. It unblushingly accused the nuclear regulatory agency of being a "white, male organization" with an inadequate understanding of South Africa's development challenges.[31] This time, however, its strategies failed—at least on paper. Finalized and passed in 1999, the new law created an independent National Nuclear Regulator (NNR) and cemented its authority to regulate radiation in mines.

Not all Chamber experts were ostriches. Denis Wymer, for example, advocated for a prompt response to the problem of radioactive mine drainage. He warned, "We have little choice but to accept that some of the radioactivity levels measured may indeed be cause for some concern." The time for secrecy had ended. Wymer recommended "a proactive and open strategy for addressing the issue" of radioactive mine water. Perhaps eyeing mounting tobacco litigation in the US, he argued that it was "only a matter of time before these results become public knowledge."[32]

Wymer was right. Everyone knew it. In 1997, the Institute for Water Quality Studies (IWQS) initiated a study aimed at establishing the radiation doses to West Rand residents from water consumption. The project's coordinating and technical committees included representatives of water suppliers, mining companies, and a variety of state agencies, along with some university researchers. Committee members diverged on how to define and study "the problem."[33] Their conflicts, which played out in negotiations over budget, technical equipment, and expertise, limited the study's reach. Ideally, for example, experts agreed that continuous monitoring over an extended period would provide the fullest picture of radioactivity levels. But the IWQS study could only fund weekly sampling during its first six months and monthly sampling during the second six months. It also had to limit its focus to the "drinking water pathway" for radiation exposure. Radioactivity in sediments, and in the food chain, would need to await future research.[34]

The report deployed its own idiosyncratic interpretation of international recommendations. Following standard scripts, it stressed that "economic and social factors" should guide all decisions about "reasonable" remediation. The ICRP had recently lowered its public radiation exposure limit from 5 to 1 millisieverts/year. But since many countries stuck to the higher limit, South Africa could legitimately follow suit, especially since uranium mine remediation elsewhere typically aimed for the 5 millisievert/year limit in the short term, leaving compliance with the 1 millisievert/year limit for the longer term.[35] The ICRP's "philosophy" justified this approach for "situations where the sources of exposure are already in place and radiation protection has to be considered retrospectively." These cases received a special bureaucratic designation: existing exposure scenarios.

In such scenarios, IWQS experts affirmed, the cost of remedial action, "including social costs, should be more than offset by the [likely] reduction in radiation dose." The scope of remedial action had to be "optimized" in order to maximize "the net benefit to society."[36] It all sounded very technical and reasonable. The generic term *society* conveniently elided the differential effects of contamination. And who could object to the supreme efficiency of "optimization"?

If you accepted these framing principles, the report had good news. Radioactivity levels did not require "immediate remediation" at any of the sampled sites. Most featured "ideal" or "acceptable" water quality.[37] Two sites with less "comforting" results simply required "further investigation."[38] In future, mines should limit their water discharges to 1 millisievert/year of radioactivity during both operation and decommissioning. And that was that. Despite absent data and methodological shortcuts, the report claimed to have "established with reasonable certainty the representative radiological status of the water resources in the catchment." Findings, insisted its authors, were sufficiently robust to undergird "a national strategy and action plan for routine and follow-up monitoring of radioactivity in public water streams, as part of an integrated approach to water quality management."[39]

Other experts harbored doubts, however. The IWQS study left several big questions unaddressed. Did radionuclides accumulate in riverine sediment? If so, could they be "remobilized" in the water, transporting radioactivity farther downstream? In 2002, a different study by the Water Research Commission (WRC) answered both questions with a clear yes.[40] Sediments could contain significant concentrations of uranium and its radionuclides, which would redissolve in streams if the water turned sufficiently acidic.[41] This had in fact occurred: uranium appeared far downstream from its entry into the waterways, including in localities "where one would not expect particulates to reach."[42] Five of the WRC's sampling sites had uranium levels above regulatory limits.[43] The IWQS had spoken way too soon. The commission recommended a "more detailed exposure assessment," starting with the spot at greatest risk: Tudor Dam. As South Africa's democratic dispensation began to find its footing, bureaucratic battles and expert disputes kicked into high gear.

2.2 Decant from a West Rand mine flowing into the Krugersdorp Game Reserve, 2011. Samantha Reinders.

Even Though Everybody Knew

The ravages of boom-and-bust capitalism didn't just happen. They weren't an unavoidable effect of the market and its allegedly autonomous behavior. Instead, these ravages resulted from human purpose, including deliberate negligence. To wit: in August 2002, the same month that the WRC released its report, rising acid waters in the West Rand mine void decanted onto the land. The timing had been predicted with remarkable accuracy by Gauteng's 1996 Strategic Water Management Plan. One of its authors later told the *Mail & Guardian* that failure by the Department of Water Affairs to take appropriate measures had given "all the mines time to get rid of their liability." Few were left to shoulder the blame when the decant began. "It caught everybody off guard, even though everybody knew about the whole thing."[44]

"Everybody," here, meant the bureaucrats and experts who populated government agencies and mining houses. Later decants would make prominent headlines, but this first one flew under the radar. Instead, in August 2002 South African media were preoccupied with the World Summit on Sustainable Development, held in Sandton, the wealthiest of Johannesburg's northern suburbs. Water was high on the summit's agenda. A huge entertainment and conference venue was re-

2.3 Opening pageantry at the 2002 World Summit on Sustainable Development. UN/Eskinder Debebe.

named the Water Dome for the occasion. Filled with booths that collectively promised to bring potable water to the world's poor, the Dome scored Nelson Mandela to deliver its opening address. The struggle hero and former president focused on progress: "When I return, as I often do, to the rural village and area of my childhood and youth, the poverty of the people and the devastation of the natural environment painfully strike me. And in that impoverishment of the natural environment, it is the absence of access to clean water that strikes most starkly. That our government has made significant progress in bringing potable water nearer to so many more people than was previously the case, I rate amongst the most important achievements of democracy in our country."[45]

I found no indication of whether Mandela knew about the decant when he spoke. His celebration of progress echoed the pageantry of the summit's opening ceremony, which featured scores of smiling South African schoolchildren (complete with their hand-drawn posters) holding hands and singing under a giant balloon of the world. It comported far less well with the twenty-five thousand protestors who marched from Alexandra township to Sandton on August 31, 2002. Organizers staged the largest demonstration since the end of apartheid, carefully choosing the protest route to highlight how "the massive unemployment, lack of essential services, housing evictions, water and electricity cut-offs, environmental degradation, and generalized poverty that is present-day Alexandra sits cheek-by-jowl with the hideous wealth and extravagance of Sandton where the W$$D is taking place."[46] Government response to the protest also belied the image of the new South Africa so carefully cultivated by Mandela during his presidency. Arrests, tear gas, and rubber bullets stood starkly at odds with the Rainbow Nation that was supposed to lead an African Renaissance. Anyone in government who hoped the demonstration was a one-off effort to capture international attention sorely misread the situation. The march turned out to be a harbinger of future action, inspiring hundreds of service delivery protests in subsequent years.[47]

Another five years would elapse before the West Rand decant made national headlines. Meanwhile, the contamination continued to spread. In 2005, the Department of Water Affairs deployed a provision of the National Water Act to order six operating mine companies to develop long-term solutions for Gauteng, but later rejected their proposal. Acid mine drainage also plagued the coalfields in Mpumalanga province, east of Gauteng. In September 2007, a broad coalition of environmental groups joined forces to call for action.

One of these conveners was Mariette Liefferink. A tall, striking, Afrikaans-speaking white woman approaching the end of middle age, Liefferink did not conform to the stereotype of an environmental activist. She rarely left the house without being dressed to the nines—often in high-necked, silk Chinese dresses, and heels, her long blonde hair impressively coiffed and her face impeccably made up. Her first foray into environmentalism began in 1995, when Shell Oil applied to build a petrol station near her then-home in Sandton. After an initial round of community opposition, her neighbors let go of the issue. Liefferink could not. She began to read up on the law, and on potential harms. She created brochures, which she printed at her own expense and distributed door to door on the weekends—a practice deeply familiar to her as a former Jehovah's Witness. But her neighbors remained passive. Shell tried to intimidate her. When she wouldn't stop, they tried to bribe her. She refused, instead spending 100,000 rands of her own funds to pursue the case. She repeatedly wrote government officials, receiving no replies. She followed up with phone calls, which they ignored.

And then something shifted. An environmental nongovernmental organization (NGO) agreed to help and arranged for her to participate in a panel at a pre-meeting of the 2002 World Summit on Sustainable

2.4 Protests outside the UN World Summit on Sustainable Development in Johannesburg in 2002. Courtesy of the International Institute for Sustainable Development/ENB-Leila Mead.

2.5 Protests outside the UN World Summit on Sustainable Development in Johannesburg in 2002. Courtesy of the International Institute for Sustainable Development/ENB-Leila Mead.

2.6 Mariette Liefferink leading a tour of mine residue areas, 2019. Boxer Ngwenya.

Development. Her presentation made a splash. The next day, Shell informed her that it would abandon its plans. "I am relieved," she told journalists. "Now I can continue with my life."[48]

But the seven-year struggle had left a strong imprint, and Liefferink's unlikely victory over Shell had put a fire in her belly. She began reading about the decant on the West Rand. The more she learned, the hotter her outrage flamed. The problem was orders of magnitude larger and more complex than that posed by a single petrol station, yet no one seemed to be calling companies or the government to account. Liefferink took up the fight, drawing on lessons learned during her previous campaign. In June 2007, she submitted a report on polluted water in the Wonderfonteinspruit catchment to a parliamentary hearing on the future of nuclear energy for South Africa.

Liefferink also got busy building a coalition. She knew she could not wage this fight on her own. Experience had taught her that organizations had more credibility than individual citizens. They also enabled fundraising. In 2007, Liefferink created the Federation for a Sustainable Environment (FSE). She worked with other NGOs to convene a national meeting of activists pursuing mine-related pollution problems. They recruited George Bizos, the famed human rights lawyer and anti-apartheid activist who'd first made his name as part of the team that defended Nelson Mandela, Govan Mbeki, and Walter Sisulu at the Rivonia trial in 1963–64.

That got the media's attention. Articles began to appear in the daily press—only a trickle at first, but by 2008 reporters realized that they had a huge story. Early headlines could seem bizarre. "Uranium Killed Cow—Autopsy Suppressed," proclaimed the *Cape Argus* in January 2008. Soon, though, journalists deepened their investigations. Liefferink seized every opportunity to help.[49] She transmitted scientific studies and government reports to the media, conducted tours of the devastated landscape, and introduced reporters to farmers and shack dwellers who testified to negative effects on their health, their animals, and their crops.

Reporters quickly discovered that the anger of West Rand residents had reached a boiling point. "The water kills everything," said white farmer Sas Coetzee. "We have a moral responsibility not to allow this to enter the food chain," added his brother Douw. The farm built by their family over three generations had become worthless. They were contemplating legal action, but "trying to fight the mines on our own is like throwing stones at a giant." Other farmers reported animals born with deformities, as well as their own ill health: sluggishness, swollen limbs, digestive troubles, strange new allergies, headaches that lasted for weeks, and more. The endless cycle of contradictory reports and broken promises deeply frustrated residents: every time a scientific study acknowledged the scope of the problem, it seemed, another would come along and dismiss the findings. Solutions were promised, then abandoned. "Our regulators are toothless and gumless."[50]

As journalists dug in, they began to understand the vast scope of the problem. Religious sects used water from polluted streams for their rituals. People smeared riverine sediment on their faces as sunscreen, or as a treatment for acne. Pregnant women continued long traditions of geophagy, eating (now contaminated) dirt to ward off ill health. Some scientists worried that the decant would seep through the Krugersdorp Game Reserve and reach the Cradle of Humankind, recently declared a World Heritage site. And all that was just in the Western Basin. Pumping in the Central Basin mine void had ceased in 2008; by 2011, some warned, the high-rises in Johannesburg's central business district might be threatened. Less grave (but comically ironic): the rising waters threatened to flood the tourist mine at Gold Reef City, a super-white mining-themed amusement park built atop an abandoned shaft.

The longer the inaction, the worse the problem became. The slow violence of racial capitalism continued to mount. In this manifestation,

2.7 A woman from Bekkersdal informal settlements washes her blankets using water from the highly toxic Donaldson Dam, 2011. Samantha Reinders.

2.8 Lydia Mokoena, a *sangoma* (traditional healer), performs a baptism in Soweto's Klip River, 2011. Acid mine drainage flows into the river, where thousands of baptisms take place each month. Samantha Reinders.

infrastructural neglect vectored the racism of its predecessor, and poor whites also suffered.

Invoking the International

South Africa was hardly alone in confronting acid mine drainage. The problem plagued most mineral extraction ventures. It sometimes even appeared in unmined areas that contained sulfide minerals, justifying less accusatory-sounding nomenclatures such as "acid rock drainage" or simply "acid drainage." During the last decades of the twentieth century, the ramp-up of environmental oversight meant that mining conglomerates operating in places with substantial regulatory capacity could no longer ignore the issue. In 1998, a coalition of ten corporations, including the South African mining giant Anglo American, formed the International Network for Acid Prevention, "dedicated to reducing liabilities associated with sulphide mine materials."[51]

Let's pause to take that in. "Liabilities." Not "harm."

The move was a classic in the residual governance repertoire: when serious threats to minimalism and incrementalism loom, corporations scramble to define best practices in the hope of influencing imminent regulations. The initiative came just as the US Environmental Protection Agency was finalizing its first handbook on Superfund mine remediation. International conferences on AMD-related topics abounded. Uranium was a frequent topic of concern at these meetings. Contrary to early Cold War fantasies, the radioactive mineral was extremely common. This meant that residues from mining other metals were frequently uraniferous. As G. Wendel might have said, there were many, many curies of rock scattered around the world's abandoned mines.

In September 2002, scarcely a month after the West Rand decant began and just two weeks after the conclusion of Johannesburg's World Summit on Sustainable Development, some three hundred scientists and engineers gathered in Freiburg, Germany, to discuss their research on uranium in the aquatic environment. A minister welcomed delegates by reminding them that Saxon engineering had shaped mining techniques the world over. German experts dominated the meeting, but Australians also put in a strong showing. So did South Africans, with an all-white delegation.[52] Delegates represented a wide spectrum of disciplines, including hydrogeology, pedology, botany, actinide chemistry, and physical geography. As in so many places, mine heritage offered a source of pride. The minister predicted that delegates would enjoy seeing the "traces of ancient mining." Not to mention "the beautiful landscape of the nearby ore mountains."[53] The aestheticization of mine ruins has long served political purposes.

The area around Freiburg had the longest known history with uranium. Agricola himself had described lung disease in the silver miners of Saxony, attributing the ailments to the inhalation of "metallic vapors." His assessment wasn't far off. Today, those "vapors" are known as an aggressive carcinogen: radon gas, a deadly decay product of uranium. Remediation efforts had begun in Germany, as well as in the US and Canada. But with five times more uraniferous waste than anywhere else, Gauteng presented a key site for study.[54]

Of the 130 papers presented at the conference, only one took the concerns of traditional landowners seriously. Its author, Peter Waggitt, had played a formative role in the Office of the Supervising Scientist, an oversight body for uranium mines on or near Aboriginal land in Australia's Northern Territory.[55] Waggitt advocated for "long-term stewardship" that included "consultation and information exchange with stakeholders."[56] His appeal doubtless seemed irrelevant to those concerned with the microdetails of uranium's behavior in aquatic environments. Or worse, dangerous: one of Waggitt's compatriots dismissed demands to reassess closed mine sites as "unfair and unreasonable" to companies.[57] In this and other ways, the Freiburg meeting exemplified a widespread pattern identified by geographers Caitlynn Beckett and Arn Keeling: even as experts present remediation plans to "stakeholders" in a show of consultation, "options are outlined only after experts have already defined the problem and the possible solutions."[58]

Geographer and chemist Frank Winde weighed in with four papers. Born in East Germany during the Cold War, his university training spanned the collapse of the Berlin Wall and German reunification. Working on water contamination from Germany's Wismut uranium mine alerted him to similar issues arising on the Rand. In 2002, he went to South Africa to pursue comparative research on waterborne uranium migration. Once there, he began to wonder: Why hadn't South Africa benefited from international remediation programs and expertise? Conversely, what lessons did the hollow Rand hold for the rest of the world? The more he learned, the wider he cast his net. His South African stay, projected for two years, would turn into twenty years (and counting).[59] In 2012, Winde became a full professor at North-West University's Vaal Campus, not far from the West Rand contamination.

At the time of the Freiburg conference, though, Winde had only just begun studying uranium in the Rand's sediment-water systems. He initially aimed to build a model to describe how uranium seeped from tailings deposits into waterways and thence to the biosphere. A variety of complex factors shaped this process: diurnal and seasonal fluctuations in the pH of waterways, photosynthesis of riverine vegetation, rainwater chemistry, and more.[60] Scientific consensus at the time optimistically held that "once adsorbed in wetlands, radionuclides and other metals . . . remain fixed in place."[61] But samples collected by Winde and his South African collaborators (who'd all been involved in the 2002 WRC study) suggested that "any change in environmental conditions could release pollutants."[62] Wetlands did not permanently trap uranium and its radionuclides. Runoff from unmaintained slimes dams, furthermore, made them "particularly prone" to AMD.[63]

Ultimately, Winde and his colleagues aimed to assess radiation exposure in Gauteng's population. As they dug deeper into the West Rand's exposure pathways, they found pollution patterns of staggering complexity. In addition to runoff from slimes dams, mine voids in the region had resulted in over one thousand sinkholes. Many of these had been "filled for stability reasons with uraniferous slimes material," which further aggravated the situation.[64] Uranium contamination in South Africa, they concluded, constituted "an environmental problem of an extraordinary dimension," not least because uranium and other mine waste particles refused to cooperate with cursory corporate plans.[65]

In sum, generalizable international expertise could only go so far. Ultimately, each ecosystem had its own history, characteristics, and human relationships. Each had to be studied on its own terms. This conclusion would not please fans of residual governance.

Manufactured Ignorance

The demand for remedial action on the Rand was building. But who bore responsibility? The few remaining mine operators argued that it shouldn't fall to the "last man standing" to carry the full burden. Corporate shuffling had enabled the largest companies to walk away from the messes they'd made in previous incarnations. National and provincial budgets already struggled to fund housing, infrastructure, public services, and health care. Entities kept passing the buck. Research continued.

In 2004, the WRC commissioned yet another report. This one was a metastudy, aimed at synthesizing existing research. Everyone agreed that the chemical properties of uranium and other metals posed risks of kidney and other organ damage. Indeed, some officials preferred to emphasize the dangers of uranium's chemotoxicity, arguing that these far exceeded the dangers of its radiotoxicity. They thus hoped to mitigate the finding that uranium levels could reach up to one thousand times background levels in the most contaminated zones of the Wonderfonteinspruit. The decant had left one lake with levels exceeding background by a factor of ten thousand.[66] Such findings required enlisting the National Nuclear Regulator—precisely the outcome most dreaded by the mining sector. The NNR didn't like it either; its personnel already had a full docket of more conventional nuclear problems to fix, and few were enthusiastic about regulating the mining sector.[67]

By now you can guess the industry's first line of defense against this threat to the comforts of residual governance: manufactured ignorance. One particularly displeased mining house persuaded the Department of Minerals and Energy to impose an embargo on the 2004 WRC report. But Mariette Liefferink refused to let the report languish in oblivion. She worked with journalists to keep the story alive. After two years of steady pressure, the report finally became public. But not before the NNR inserted a disclaimer distancing itself from the findings: "The NNR makes use of an internationally recognized methodology as well as international norms and standards in its radiological risk assessment of uranium levels in water. The methodology used by the WRC in this report is inconsistent with these norms and standards. Its research essentially assesses the chemical risk of uranium in the water body. In the circumstances, the NNR is not in a position to concur with the methodology and conclusion of this report."[68] Invoking the authority of international recognition, the NNR stressed the distinction between uranium's chemical and radiological hazards. The former were banal, and not within the agency's purview. The latter lay squarely within its domain, and as such the NNR refused to countenance input from other government agencies. Instead, it commissioned BS Associates, a German consulting firm with offices in South Africa, to conduct a radiological risk assessment of the catchment.

Private consulting firms, often run by former industry insiders, were significant players in what Mpofu-Walsh calls the privatization of apartheid. (You'll encounter several more examples in subsequent chapters.) The NNR could reasonably hope that delegating the risk assessment to consultants would result in lower risk estimates. But the men at BS Associates did just the opposite. The Brenk report, as their study became known, found that—contrary to assumptions that drinking water destined for humans offered the most significant pollution pathway—three other pathways posed far greater dangers: crops and pastures irrigated by contaminated water, cattle imbibing contaminated sediments at riverbanks, and agricultural land contaminated by runoff from slimes dams. Radiation exposure at half of the sampled sites exceeded the 1 millisievert limit for public exposure. Some sites were ten times that, and a few exceeded the limit by two orders of magnitude. With as much diplomacy as they could muster, the consultants concluded that "the mollifying results of [earlier] radiological impact assessments are not confirmed and are—from our point of view—not appropriate."[69]

The Brenk report clarified the complexity of contamination. And of its measurement: actual exposures from each pathway depended on age, livelihood, habits, time of year, even time of day. Precise quantification proved impossible. Infant exposure offered a case in point: "Dose assessments for infants would require considerations of the exposure pathway 'breast milk feeding,' . . . a relatively complex pathway that necessitates a transfer factor from the mother's uptake of radionuclides to breast milk and representative data for the rate and duration of nursing. For German habits these data are available. . . . However, we feel that lactation in South Africa could be much different."[70] South African licensing guidelines did separate population groups by age. But the categories began at age one. Infants literally didn't figure into the equations.

Another complication arose from the practice of using "reference persons" in modeling exposure pathways.[71] This entailed assuming "standardized habit parameters" (eating, drinking, mobility) for people in a given age group. The consultants warned, however, that "actual habits of residents and critical groups" could differ significantly from that of a generic "reference person." Standardized exposure times wouldn't apply to "the jobless persons living at the bank of the Padda Dam from fishing." Data on bioaccumulation and biomagnification of contaminants in fish were still lacking; these mattered because "jobless persons" and their families who relied on the dam for protein probably consumed more than the 25 kilos of fish assumed by licensing guidelines for South African adults.[72] In these and other respects, the German authors proved more attentive to the precarious living conditions of area

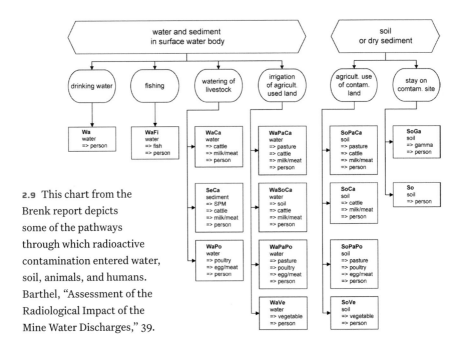

2.9 This chart from the Brenk report depicts some of the pathways through which radioactive contamination entered water, soil, animals, and humans. Barthel, "Assessment of the Radiological Impact of the Mine Water Discharges," 39.

residents than their South African interlocutors were. Or at least more willing to take those conditions seriously when estimating exposures.

Bottom line: studies claiming that radiological exposures in the Wonderfontein catchment fell below the international limits were wrong. Exposures significantly exceeded those limits in at least half of the sites sampled. The consultants presented twenty-six recommendations for future action. These included remediating the worst sources of pollution, restricting public use of contaminated sites, implementing a comprehensive monitoring program, and conducting more robust data collection with improved research protocols. All of which would cost money, of course. Would the polluters pay?

These conclusions embarrassed the NNR. Rather than publicizing them, the nuclear regulator quietly issued a directive to mines operating in the catchment: residents should be barred from relying on Wonderfonteinspruit water for fish, crops, and livestock. The NNR refused to share a copy of the directive with municipal authorities. (Score another point for manufactured ignorance.) But when Harmony Gold issued a public proclamation prohibiting use of the spring water, the secret leaked out. Earthlife Africa immediately protested: Harmony did not have the power to take away access to water. By failing to provide either clean water or adequate compensation for harm and loss, the directive violated citizens' constitutional right to "sufficient water."[73] Interdictions looked good on paper and could potentially limit liability. But they did nothing to ameliorate actual living conditions.

Of Cabbages and Lions

Mounting publicity increased the political urgency of remediation. Over 400,000 people lived along the Wonderfonteinspruit, in informal settlements, farms, townships, and golf estates.[74] The "state knew about [the] danger for 40 years," sighed the *Sowetan*, but "bureaucrats and their political masters in democratic South Africa have been [just] as unwilling to offend the [mining] magnates and their investors as their counterparts were during apartheid." As a result, "unwitting communities—megarich and pitifully poor—drink, bathe and play at unfenced poisonous sites."[75] The Afrikaans-language paper *Beeld* also raised the alarm, especially after the Brenk report's lead author was mysteriously blocked from delivering long-scheduled talks at a 2007 Environmin conference.

As news reports proliferated under increasingly sensational headlines, the public had trouble knowing what to believe. Should they really expect a "toxic tsunami," as one headline proclaimed? Did farmers need to shoot the thousands of cattle who'd drunk polluted water, as another warned? Studies proliferated, but some used questionable methods and created further confusion.[76] Public officials made declarations that contradicted or misstated the results of the Brenk study. Lindiwe Hendricks, the minister of Water Affairs and Forestry, told Parliament that despite radiological contamination of sediments along the stream, the water itself did not exceed NNR limits and was "therefore safe for drinking purposes." Despite his agency's directive enjoining mine companies to prohibit water use, NNR head Maurice Magugumela insisted that the public had "no cause for concern."[77]

If neither scientists nor the ANC-led government could be trusted to stand up to the mining companies, where could citizens turn? A variety of public interest groups formed, including the Wonderfontein Action Group and the Public Environmental Arbiters. Meanwhile, the Federation for a Sustainable Environment redoubled its work. Liefferink sought new alliances, collected data and documents, attended meet-

2.10 Radioactive pollution map of the Upper Wonderfonteinspruit. Red indicates elevated levels of radiation, and blue indicates lower (but still measurable) levels. Du Toit, "Background Report on Communities at Risk within Mogale City," 50.

2.11 Radiometric count of the area around the West Rand Gold Mine in Kagiso. Red indicates elevated levels of radiation, and blue indicates lower (but still measurable) levels. Du Toit, "Background Report on Communities at Risk within Mogale City," 52.

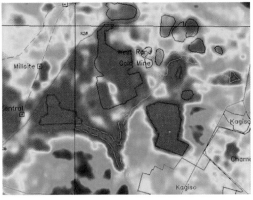

ings all over the region, and maintained her active relationship with the press.[78]

Mounting public concern prompted the Mogale City Local Municipality, home to much of the catchment's pollution, to commission an independent assessment from its own environmental manager, J. Stephan du Toit.[79] How did radiological and other toxic contamination affect the municipality's most vulnerable denizens? To answer this question, du Toit triangulated findings from existing studies with GIS data and his own research. His overview maps of the whole plateau depicted high-risk areas. The West Rand consistently featured elevated radiation levels, though hot spots existed elsewhere too. Detailed aerial maps portrayed radioactive contamination in the Mogale City jurisdiction.

Du Toit documented the use of surface water by small-scale farmers, who reported illness in themselves and their employees, decreased fertility in cattle, and other problems. But he could not find robust enough research to bolster their claims. One researcher at the University of Johannesburg had analyzed a cabbage from the catchment and found highly elevated levels of several metals. The Krugersdorp Game Reserve, also in the catchment, reported that several of its lions had died "under suspicious circumstances," which postmortems attributed to the ingestion of heavy metals and sulfates.[80]

Still, one cabbage and a few dead lions did not suffice. More data were needed to document "the potential bio-accumulation of pollutants in plant and animal tissue and the concentration of these pollutants up the food chain."[81] Food safety required urgent attention. As du Toit pointed out, this wasn't exactly a new conclusion. Earlier reports had already noted that residents of informal settlements had no choice but to use "contaminated ground and stream water for personal hygiene and drinking. With above-average infection rates of HIV/AIDS and chronic and acute malnutrition, this subpopulation is particularly vulnerable to additional stress on the immune system by contaminants such as uranium."[82] Du Toit counted some eighty-one thousand township residents, plus an unknown number of shack dwellers, who faced serious risks from contaminated water.

Du Toit had little patience with the NNR's prevarications. He dismissed the agency's concern that publicizing radiological impacts would create "public hysteria," noting bluntly that "failure to inform the public can prove fatal."[83] Withholding information, along with the "sedulous avoidance" of epidemiological research into human contamination, would lead the public to infer "active concealment of truth."[84] Du Toit called for a "ground-truthing" exercise to validate his conclusions, along with a "comprehensive population study" to assess the health impacts of heavy metal poisoning and radiation exposure.

This plainspokenness contrasted sharply with the circuitous language of most official writing. Du Toit later explained to Liefferink that they'd discovered heavy metals in his young son. The family had been living in the Krugersdorp Game Reserve when "[Hannes] was diagnosed with ADHD at the age of 5 years. He also had elevated levels of mercury, cadmium, aluminium and lead. It was later determined by the Council for Geosciences (for both my daughter Christi-Ann and Hannes) that the lead accumulated in their systems were originating from Uranium (based on the lead isotope ratios). We intensely assessed the various pollution sources (point & diffuse) that they were confronted with and the various pollution pathways . . . highlighting again the need for a high confidence population health study."[85] Du Toit had skin in the game. His vehemence was visceral.

Du Toit's report circulated through Mogale City's municipal bureaucracy, apparently without much fanfare. The town's Sustainable Development Facilitation Forum endorsed its conclusions. But then the report disappeared. Another four years would elapse before it once again snagged the attention of the municipality's mayor and executive committee.[86] Meanwhile, it circulated, in the form of an unpublished Word document, among researchers and activists. My own copy came directly from Liefferink, who generously shared hundreds of such documents.

Mounting criticism led the Department of Water Affairs and Forestry to launch an initiative in late 2007 to rehabilitate the catchment in partnership with the NNR. Each agency limited itself to its respective domains of expertise. The effort focused on water pollution pathways and the radiotoxicity of uranium. It did not address uranium's chemotoxicity, nor did it consider any other heavy metals. Perhaps strangest of all, it permitted the use of data from only three studies, which together covered a scant one hundred sampling sites along the stream.[87] Areas outside the catchment were ignored. So were spots between sampled sites within the catchment.[88] The agencies called on a team of South African, Canadian, and American experts to establish research protocols.

The team tasked Frank Winde with producing a hot spot map to prioritize sites for intervention. His approach combined uranium levels in water and sediment with the probability of exposure, leaving enough leeway to "adjust priorities according to political preferences." The map could thus inform different remediation benchmarks (natural background versus regulatory limits), regulatory limits (international versus South African), and uranium concentration levels (average levels versus maximal ones). Stakeholders approved. But the larger rehabilitation effort stalled for lack of funding. The foreign members of the team met only once. A consultant originally contracted to handle mundane administrative tasks took things over and compiled a Remediation Action Plan in 2008.[89] His lack of expertise resulted in a deeply flawed plan, which no one accepted: not the experts, not the mining industry, and certainly not residents. Scientists objected to the plan's narrow mandate. Stakeholders felt their perspectives and interests were excluded. Winde felt the administrative consultant had hijacked the project.

More than three years and "significant funds" later, remediation efforts went back to square one.[90] Meanwhile, the vast scope of the situation became increasingly visible. After scrutinizing data from over a dozen studies, Winde estimated that some 2,200 tons of uranium had accumulated in the fluvial sediments in the Wonderfonteinspruit catchment as a result of mining projects conducted over the previous 122 years. While cities such as Johannesburg and Ekurhuleni drew water from the Vaal Dam and the Lesotho Highlands Water Project, many West Rand municipalities sourced their water from local streams such as the Wonderfonteinspruit and the Tweelopiespruit, which were loaded with uranium. A wicked problem indeed, one well on its way to super wickedness.

The Greenest District in South Africa?

In the absence of a plan, the question of what counted as remediation remained open. Whether or not they admitted it, experts, state bureaucrats, and mine officials knew that eliminating all contamination and returning the land to pre-mining condition would be impossible. What should be the scope of remediation? What should be prioritized? How much was enough—and who would decide? Who were the stakeholders, and how should they be included in planning and monitoring? Could the molecular approach—treating sources of contamination separately so that bureaucrats and experts could stay on their specialized turfs—gain any public legitimacy? Further delays seemed politically untenable, but funding restrictions limited data collection. Was there enough to inform decision making? Who would pay?

All of these questions required reckoning with regulatory standards. Here too, South Africa floundered. Frequent changes in international standards didn't help. In 1971, the World Health Organization (WHO) had tentatively identified the need to limit uranium in drinking water, but didn't have enough data to establish a cap. When the WHO next published drinking water guidelines in 1984, uranium had dropped off the radar again. The element popped back up in 1993, but in the continued absence of robust chemotoxicity research, the WHO relied on radiological data to guide its standard, setting the cap at 140 micrograms per liter (μg/l) of uranium. A 1998 revision reduced this dramatically to 2 μg/l, but declared the limit provisional because existing studies weren't conclusive; the limit could prove "difficult to achieve in areas with high natural uranium levels."[91] Terminology facilitated slippage: nuclear experts called ore extracted from the ground "natural uranium," in contrast to the "enriched uranium" treated for use in weapons and reactors. The nomenclature had long enabled operators to downplay radiation from mining as equivalent to "natural" background.[92]

Regulatory standards are slippery beasts. They appear to protect health and environment by limiting permissible contamination. And there's no doubt that without regulation, the planet would be in even worse shape. But standards are also—always—the product of political negotiation. Corporations and their allies bring far more resources to the negotiating table than communities can. Too often these stark power imbalances result not in meaningful restraints, but rather in perfunctory regulations that effectively function as permits to pollute (as long as you don't pollute too much).

This dynamic is exacerbated by the weak authority of international organizations, which can neither mandate nor enforce limits. They can only make recommendations. It's up to member nations to adopt, change, or ignore those values. Bear in mind, too, that international organizations are staffed by experts from member nations, many of whom have their own vested interests in the industries they purport to assess impartially. Consider what happened after the WHO raised its cap for uranium in water to 15 μg/l in 2004. Germany stuck with the earlier, more restrictive 2 μg/l recommendation. Other active uranium producers had caps higher than the WHO guidelines. But apartheid South Africa had been off the charts, deeming a whopping 8,000 μg/l to be low risk. Talk about residual governance! Not until 2011 did South Africa align its limit with the WHO value of 15 μg/l. "Given that all guideline values ultimately aim to protect universally identical human health," Winde remarked wryly, "it is difficult to scientifically explain the large differences in acceptable U-levels by relying on medical arguments."[93]

Winde found it particularly odd that the WHO cap had gone up rather than down in 2004. He speculated that pressure by the International Atomic Energy Agency (IAEA) had played a role. The increase played into excitement about a "nuclear renaissance" that drove up the price of uranium.[94] Winde's speculation rested on solid historical precedent. The IAEA had been founded to promote atomic energy, not restrict it. The agency had a long track record of diminishing nuclear risks. Most notoriously, it dramatically downplayed morbidity and mortality from the Chernobyl accident: where outside cancer specialists attributed some fifty thousand deaths and illnesses to the accident, the IAEA insisted the number didn't exceed four thousand.[95]

The chaos and contradictions of standard setting are enhanced by a popular industry playbook that invokes scientific uncertainty to argue that tough regulations needlessly impede growth. Tobacco lobbyists, the oil industry and its climate change deniers, asbestos producers, and the nuclear industry all insist that without studies conclusively proving the ill effects of a specific exposure, implementing regulatory benchmarks requires more research.[96] They further insist that proper research involves isolating individual contaminants to determine precisely which ones cause ill effects. Such stipulations undergird what historian Evan Hepler-Smith calls "molecular bureaucracy": institutions, regulations, and knowledge production that deal with toxins molecule by molecule, element by element.[97] This compartmentalized approach makes it extremely difficult to apprehend synergistic effects of multiple forms of exposure. It leaves no room for embodied or experiential knowledge, summarily dismissed as anecdotal evidence.

Countering these structures has always been extraordinarily difficult and time consuming. But Liefferink was fearsomely persistent. The Jehovah's Witnesses had taught her to cope with rejection. Between 2007 and 2010, she prepared 13 submissions to parliamentary bodies, the Public Protector, and the Human Rights Commission; attended 84 academic symposia; engaged in 252 workshops and site visits; appeared 750 times in the media; and distributed 16,800 brochures and questionnaires.[98]

In November 2011, a member of the Dutch Reformed Church in Helikon Park sent Liefferink a long list of ailments reported by her congregation over the previous five years. Most of these involved some form of malignancy: brain tumors, lung cancers, lymphomas, and leukemias. Kidney and liver problems also abounded. Several reports concerned young children who had died at age four, or six, or eleven. Some ailments were downright bizarre: "A lady was diagnosed with a hole in the eye. A neighbor of Mrs. Coreeges has the same problem."[99] Acutely aware that residents in other parts of the Rand also suffered, Liefferink continued to broaden her investigations.

Black and Coloured citizens experienced their contaminated surroundings not just as a source of physical misery, but also as a betrayal of liberation promises. The new South African constitution guaranteed them the right to a clean environment. New democratic structures and processes, furthermore, were supposed to give them a voice in shaping their material and social environments. The 1998 National Environmental Management Act devoted a chapter to "integrated environmental management," an approach aimed at formally addressing the complexity of wicked problems. Its companion, the 1998 National Water Act, established nineteen Water Management Areas, each managed by a catchment agency obligated to solicit input from stakeholders via quarterly forums on the development and implementation of water management strategies.

New legislative structures required new bureaucratic devices. One such device: Environmental Management Frameworks (EMFs).[100] Intended to operate at political scales ranging from the district municipality to the province, EMFs established an area's aspirations for land use, housing, access to services, and commercial and agricultural development. They also produced plans for achieving these aspirations in environmentally and historically sensitive ways; honoring heritage was part of their mandate. Draft frameworks had to respond constructively to issues raised during public meetings.[101] In principle, EMFs that made it all the way through the approval process would be officially "gazetted," making them enforceable in a court of law.

In short, administrative mechanisms for consultation and repair appeared to be firmly in place. So why did mine residues still permeate the lives of those who lived nearby? And why did so many Gauteng residents still feel betrayed?

Consider the West Rand District Municipality (WRDM). Home to the region's most polluted areas, the WRDM stretched out over 2,400 square kilometers and comprised four local municipalities (including Mogale City, where I'll take you in chapter 4).[102] A slice of the Cradle of Humankind World Heritage Site also fell within its ambit. The 2011 census counted close to 821,000 inhabitants. Migration patterns kept shifting, making the housing challenge even more complex. In 2001, 34 percent of WRDM residents lived in informal settlements. By 2007, that proportion had increased to 42 percent. An influx of people seeking job opportunities exacerbated the "housing backlogs."[103]

The WRDM didn't have the capacity or resources to produce an Environmental Management Framework on its own. It contracted the job to BKS, an engineering consultancy founded in 1965 (the heyday of apartheid). The firm offered the West Rand a noble vision of its future: "to be the greenest district in South Africa and to provide an African example that Sustainable Development is not just a good choice, it is the best choice."[104] (Such lofty language permeates the firm's self-presentation; the company aims to become "the African world-class leader in the art of supplying sustainable development solutions to help eradicate poverty.")[105]

The resulting EMF proposed to provide "a better life for all" by addressing acid mine drainage, promoting small-scale farming, managing waste disposal, and of course meeting the housing backlog.[106] "Environmental management," the framework concluded, "must place people and their needs first, serving their physical, psychological, developmental, cultural, and social interests equitably." This meant honoring the environmental rights articulated in the constitution. It meant equitable access to resources, and widespread public participation in environmental governance. It meant coordinating action among all the relevant government bodies. And it meant that "sensitive, vulnerable, highly dynamic or stressed ecosystems require specific attention in management and planning procedures, especially if subject to significant human resources usage and development pressure."[107]

All of which was far easier said than done. BKS flatly admitted that "the desired state of the environment for the WRDM is an ideal (utopia)."[108] Low participation in public meetings did not bode well for democratic management. Over five hundred residents had officially registered as stakeholders, but meetings rarely saw more than a handful of attendees. Mariette Liefferink attributed low attendance to a combination of factors. "Indigent communities," she noted, "often do not have the financial resources to attend public participation meetings," particularly if they have to pay for transport to attend. Relying on email to publicize the meetings compounded the problem. And "because comments from the public have in the past not been given the necessary consideration and weighting, the public perceives [the] participation process as a sham and tokenism."[109] Last but by no means least: spinning up on the technoscientific language of the reports took tremendous time and resources, which most residents simply didn't have.

Although it had dutifully jumped through all the formal hoops, the West Rand's 2013 EMF was never gazetted. So it didn't carry the force of law. Instead, it was superseded by Gauteng's provincial-scale EMF, produced by yet another consultancy the following year. The provincial EMF, however, was so superficial that it had little value as a concrete planning document.[110]

Nevertheless, Liefferink herself pressed on. She regularly attended meetings for four catchment forums, as well as some fifteen other committees concerned with environmental management strategies. To the evident irritation of some officials, she went through scientific reports and other documents with a fine-toothed comb, taking as much airtime as she needed to flag errors and question conclusions. In partnership with other civil society groups, such as the Centre for Environmental Rights, she hounded mining companies and government departments to release reports that they tried to keep confidential. She insisted that they attend to the uneven experience of pollution: the differences among water sources, the specific challenges faced by informal settlements, the details of water treatment processes, and more. Her mastery of the details, her quasi-superhuman energy, and her unfailing politeness made it difficult to dismiss her.

Treating the Toxic Tide

By 2010, demands for action on AMD had reached a boiling point. Addressing the ongoing decant in the Western Basin required more than just research. Dire predictions of imminent decant in the considerably larger Central Basin, furthermore, made that zone another urgent priority. The national government finally formed an Inter-Ministerial Committee on acid mine drainage. Members included representatives from the ministries of Mineral Resources, Water Affairs, Science and Technology, and the National Planning Commission. The committee's high-level objectives were telling: "be informed by facts and science; ensure consistency and calmness in explaining what is happening; communicate in order to regain and maintain the trust of the people."[111] Acid mine drainage had become a focal point in a wider crisis of democratic governance and legitimacy.

The committee appointed a "Team of Experts" to synthesize existing knowledge and recommend solutions. Their report, delivered to the Cabinet of Ministers in February 2011, nixed the notion that action couldn't proceed without more research. The Water Research Commission alone had produced fifty-four studies. Other entities had also proved prolific. "This has resulted in a sound but generic understanding of the process and the various components of the AMD problem in South Africa," wrote the experts, adding that the "complexity of the host and receiving environments . . . militates against a single or 'one size fits all' solution."[112] The simplifications of residual governance simply couldn't handle the wicked wastes of racial capitalism.

State agencies had developed a suite of regulatory guidelines on water protection and mine closure, as well as a strategic plan "aimed at addressing the liability of government for the thousands of derelict and ownerless mines." The *Global Acid Rock Drainage Guide*, "to which South African organisations have contributed significantly," provided "best practice guidelines"—which once again represented an industry attempt to decisively shape state regulations. In this case, the guide enabled South African state experts to argue that "sufficient information exists to be able to make informed decisions regarding the origins of the mine water, potential impacts, management strategies, treatment technologies, etc."[113]

No more dithering. Intervention had become "a matter of urgency." The decant rate in the West Basin peaked at some 60 megaliters per day during summer rains; in other seasons, it averaged 15 to 20 megaliters per day. Water was also rising through the voids in the other Rand basins. Other mining zones in Mpumalanga and Limpopo provinces also warranted close attention.

The experts presented recommendations for short-, medium-, and long-term measures. First among these: the construction of water neutralization plants.[114] Others included reducing water ingress into the voids, maintaining pumping to prevent decant, a "multi-institutional" monitoring system, and more. The experts considered a variety of active, passive, and in situ treatment technologies, repeatedly emphasizing that the choice of method depended on highly localized conditions.

Whatever the solution, the pH of the water needed to be raised. And that was expensive. Experts predicted "a shortfall between the cost of clean water produced in a plant and the revenue recoverable from the sale of water."[115] Treatment plants couldn't even cover their own costs, let alone turn a profit. So who would pay? The Department of Water Affairs had determined that 90 percent of the costs of treating acidified water on the West Rand should be allocated to three mining companies, leaving the state to cover the rest.[116] Public-private partnerships would build or expand treatment plants for all three basins, starting in the West. With flooding looming, the plant was exempted from undergoing an environmental impact assessment and heralded as a success.[117] After a decade of dithering, it certainly represented a significant achievement, raising the pH of the water it treated from the high 2s to around 6.

2.12 Western Basin AMD Treatment Facility, 2022. Luka Edwards Hecht.

In the grand scheme of things, however, the treatment plant offered a bandage rather than a long-term solution. Mine water didn't just need chemical neutralization. It also required technopolitical neutralization. And that meant potability. Water released by the treatment plant still bore significant sulfate loads. These salts left the water unsuitable for irrigation, not to mention undrinkable. The obvious solution—further dilution—would threaten the region's scarce water supply. And while a relatively neutral pH allowed many metals to precipitate out, uranium wasn't like the others. It could still travel in treated water.[118]

Besides, acid mine drainage was just one of many challenges posed by the residues of mining. Dust proved even more recalcitrant.

1910 = 2010

3

THE INSIDE-OUT RAND

WATER WAS NOT THE ONLY underground substance demanding evacuation before mining could proceed. Even more fundamentally, shaft sinking and tunneling required rock removal. The lower the ore grade, the higher the proportion of sterile stone. Once surfaced, the barren boulders had to go somewhere. Outside the shafts, the rocks piled up inexorably. Mining made mountains as well as warrens. The hollow Rand was also, simultaneously, the inside-out Rand.

With each passing decade, the mine dumps grew higher, proliferating across the plateau and reshaping its topography. Maximizing rapid profits meant working the highest-grade deposits first. In 1900, obtaining 20 grams of gold entailed hauling one ton of rock to the surface. By 1980, the quantity of residue generated by 20 grams had doubled to two tons. Some dumps consisted primarily of rock, others of sand. Many had both. As the chemical processes for ore extraction changed, so did the morphology of residues. Particles grew finer, and companies started storing the excess slurry from cyanide treatment plants in giant slimes dams: flat-topped, engineered expanses visible from space.

Regardless of form or location, turning the Rand inside out generated colossal clouds of dust all along the vertical axis, from the deepest underground shafts to the highest sand dump. From the moment the first blasts broke rock in 1886, fine-grained silica particles hung in the air, visible and palpable. It didn't take long for the inside-out Rand to infiltrate bodies, for mine workers to develop wracking coughs that made the ill effects of inhalation impossible to ignore. Rock residue scarred miners' lungs, transforming them into silicotic sacks highly vulnerable to tuberculosis. By 1902 the situation had ballooned into a crisis of such severity that the government was compelled to create a commission to investigate. Industry resistance meant that another decade would pass before the first workplace legislation forced mines to adopt mitigation measures: ventilation, dust extraction, compulsory medical surveillance, and more. South African mining magnates lost no opportunity to congratulate themselves for being world pioneers on this front, even though the measures came as piecemeal, minimal responses to ongoing pressure from the white labor union.[1]

Aboveground dust, mine companies hoped, would simply dissipate into the air. But with hundreds of thousands of miners working the seams, the mountains rose so rapidly that the winds couldn't disperse the residues fast enough, or far enough. Dumps dominated the landscape and defined the contours of city planning: Black housing downwind, white housing upwind. After 1948, apartheid took such spatialized racism to new lows. Dumps, dust, and discrimination became deeply entangled. Nearly three decades after the advent of democracy, these entanglements remain coded into the new apartheid, into the "algorithm running in the background of South African society."[2]

Until it gushed onto the surface, acid mine drainage remained invisible to all but a few officials, experts, and farmers. Dumps and their dust, however, were hypervisible from the start. Artists, journalists, photographers, and writers found the odd yellow mountains morbidly captivating. As the dumps increasingly defined the physical contours of Johannesburg, a range of representations inscribed them into the imaginations and identities of urbanites. Photographs proved particularly useful for revealing the residuality of their governance. Scientists also tackled the dumps, of course: many in service of the industry, some in service of (what was not yet known as) environmental justice.

In short, the inside-out Rand was anything but static. Its components circulated. Its materiality morphed. Its meanings multiplied.

Outside-In, Underground

In colonial South Africa, dust mitigation measures were racist by design, a primary tool through which racial capitalism etched itself onto Black male bodies. Starting in 1916, white workers, who made up just 10 percent of the workforce, benefited from medical examinations conducted by the Miners Phthisis Bureau. Black workers, exposed to far higher dust levels, were summarily handled by mine medical services. White workers benefited from compensation if their exit exams showed disease. Black workers did not. For decades, in fact, they didn't even get the exit X-ray that would have shown compensable disease. In 1930, Alexander Orenstein, one of the industry's favorite medical experts, asserted that "any attempt to make the examination of Blacks standardised on a basis of white experience [by using X-rays] would be a horrible mistake."[3] White workers had access to sanatoria for treatment. Black workers were sent home to rural areas with little to no medical infrastructure, where the tuberculosis they'd contracted on the mines, often misdiagnosed as pneumonia, spread to their family caregivers and neighbors. Asked about the likelihood of such infections, Orenstein responded flippantly, "There is very little risk, as such men soon died."[4]

Historian Jock McCulloch—who spent his career studying occupational disease across the South African mining industry before succumbing to mesothelioma himself—offered stark and succinct analysis: "In political terms, the major issue for employers and the state was *where* miners died."[5] Those who died at home did so away from the white political gaze, enabling industry to pretend that these mine workers died "soon" of diseases that everyone knew took years to kill you. Industry's favorite expert proved equally adept at deflecting scrutiny abroad, using data on white workers' health outcomes to aver that South African mines had nearly eradicated silicosis. Had anyone bothered to ask them, the women who took care of their disabled husbands back in (present-day) Botswana, Lesotho, and the Eastern Cape could have provided mountains of evidence to contradict such claims.[6]

White organized labor further entrenched the social and corporeal order wrought by racial capitalism. The all-white Mine Workers Union was completely committed to maintaining the "colour bar" that kept Black workers in subordinate positions and limited their wages. They knew that the 12:1 wage differential enabled companies to compensate sick whites while still turning hefty profits. The ledgers didn't lie; white health and prosperity came at the direct expense of Black bodies. The outcome was textbook residual governance: management of residues (silica dust) in minimal fashion (for white workers only) that depended on treating large swaths of the population (Black workers) as residual.

In South Africa as elsewhere, racial capitalism both depended on and perpetuated residual governance.

Neither Black miners nor their political leaders needed whites to understand the dangers of underground labor. Sol Plaatje, the first secretary of the ANC, wrote in 1914 of the "two hundred thousand subterranean heroes who, by day and by night, for a mere pittance lay down their lives to the familiar 'fall of rock' and who, at deep levels, ranging from 1,000 to 3,000 feet in the bowels of the earth, sacrifice their lungs to the rock dust which develops miners' phthisis and pneumonia." After years of suppression, Black union organizers finally gained traction in the early 1940s. On August 12, 1946, tens of thousands of Black workers across the Rand went on strike to demand better wages and working conditions. They knew they would be met with violence. But what was the alternative? "When I think of how we left our homes in the reserves, our children naked and starving, we have nothing more to say. Every man must agree to strike on 12 August. It is better to die than go back with empty hands," declared one man. Besides, said another, "we on the mines are dead men already." The militarized police responded every bit as viciously as they'd anticipated.[7]

The strike inaugurated a new phase in South Africa's liberation struggle. More would follow. But the dusty conditions continued. The year 1982 saw the founding of the National Union of Mineworkers (NUM) to represent mostly Black workers, led by Cyril Ramaphosa (who thirty-six years later would become South Africa's fifth democratically elected president). In 1987, the NUM launched South Africa's largest strike to date. Some 360,000 gold and coal mine workers lay down their tools to protest working conditions, wages, and the color bar.

Not until 1995, shortly after the advent of majority rule, did the umpteenth state commission on health and safety in the mines officially declare that dust levels in the mines, and corresponding disease levels, had been unacceptably high for decades. These and related findings underpinned a stack of class action lawsuits against the remaining companies. Seen as long overdue, the suits sparked the 2018 documentary *Dying for Gold*.[8] In July 2019, after more than a decade of litigation, the Gauteng High Court approved a 5-billion-rand settlement, to be paid out to silicosis sufferers by the Tshiamiso Trust. In Setswana, *tshiamiso* means "to make good" or "to correct"; the Trust website asserts its vision "to live up to our name."[9]

In theory, the settlement represented a major victory for those whose lungs were shattered by the particulate residues of their labor. But observers immediately voiced concerns about implementation.[10] Former workers have expressed deep frustration with the slow pace of payments. Mzawubalekwa Diya, who worked for Sibanye-Stillwater for twenty-seven years, was among the thousands who migrated to the mines from Bizana (Eastern Cape Province). He described his experience of the compensation ordeal to *New Frame* in April 2021: "I was told that my lungs are damaged by the silica dust at the mine. . . . Only one lung is functioning. It has been a long time. We have been going up and down with interviews and doctor's appointments, but the money is not coming. I can't afford to do anything. The Trust is very slow and they must understand that the money in their bank account is not theirs, this is the money we worked hard for as South African miners. Can they just pay me before I die?" Since then, the pace has picked up. Still, applying for compensation isn't only a matter of walking into an office and filling out paperwork. It's a nine-step process, involving a long chain of meetings and medical appointments. The Trust warns applicants that "each claim can take up to 180 days to be finalized from date of lodgement."[11] Some mine workers who've been through that ordeal find the sums they receive risible. In September 2021, for example, one former miner, speaking on condition of anonymity, called the 70,000 rands he'd received a "joke." (In 2021, 70,000 rands equaled 4,400 US dollars or 4,000 euros.) He was heading to the Tshiamiso Trust offices for an explanation. "For all the years and damage to my lungs, is this amount all I get?"[12] Establishing compensation eligibility, in short, was but one step. Civil society campaigns (such as Justice for Miners) and attorneys (most notably Richard Spoor Inc., which also played a major role in asbestos litigation), along with miners themselves, still had to fight to get former mine workers their due.

The many people who waged these battles approached underground dust primarily as a labor problem—one taken on through workplace struggle, mine medicine, union advocacy, and (eventually) compensation schemes. This meant that, complex as it was, the problem was relatively circumscribed, and the avenues for its politics clearly defined. Aboveground dust was less technopolitically tractable.

These Pyramids Speak Lungs

At least until the late 1950s, mining magnates liked to celebrate the size of their dumps. Companies proudly displayed their monumental achievements in lavishly illustrated industry journals. The dumps appeared as things of beauty, majestic "man-made mountains," visual indexes of productivity and profitability.

The monumental idiom of industry magazines contrasted sharply with the intimate portraits offered by Ernest Cole, a freelance photographer for *Drum* magazine. His 1967 *House of Bondage* presented audiences in Europe and North America with a tableau of Black life under apartheid.

Cole's work was not the first glimpse into apartheid South Africa offered to foreign audiences. In September 1950, *Life* magazine featured a fifteen-page spread called "South Africa and Its Problem." Nestled among ads depicting white women fawning over the televisions, bedroom sets, and girdles that defined white middle-class consumption in postwar America, the piece purported to "explore South Africa's great issue and dilemma, the black problem." (Not, you'll note, the white problem—the audience, after all, was Jim Crow America.) The article likened "the root of the problem" to insectoid hiving: "Blacks must work for the whites to live, and the whites must always live dependent on the black millions swarming among them." It included an excerpt from Alan Paton's novel *Cry the Beloved Country* to illustrate the dilemma of a well-meaning white man. The only Black African quoted in the piece was the unnamed and unpictured "resentful Native" who'd chalked "GOD IS BLACK" on a wall in downtown Johannesburg.

The *Life* spread featured images by the white American industrial photographer Margaret Bourke-White. During her five months documenting apartheid South Africa, Bourke-White persuaded reluctant mine managers to let her snap pictures of mine workers underground. She aimed to humanize her subjects: although marred by casually racist caption writers, many of her portraits conveyed the strength, dignity, and masculinity of their subjects.[13] She'd also been struck by the dumps:

3.1 Ernest Cole, self-portrait. Ernest Cole/Magnum Photos.

a two-page black-and-white spread near the start of the piece depicted "an appalling yellow wasteland of mountainous refuse dumps—the dusty leftovers of six prosperous decades of gold mining. . . . Nothing will grow on these or the scores of others like them which are spread across more than 50 miles of the veld." In this rendition, the dumps were above all an aesthetic affront.

Bourke-White approached her task with seriousness and concern, gaining remarkable access to a wide variety of communities. Still, her vision was that of a white American visitor. How could it be otherwise? Ernest Cole, by contrast, reached deep into spaces, experiences, and emotions beyond white gazes.

Born in 1940 as Ernest Levi Tsoloane Kole, by the age of eighteen the slender photographer was buzzing all over Johannesburg and its townships on his Vespa. Colleagues remember the young man's austerity: a devout Catholic, he never drank, never smoked, and ate with utmost frugality. "He was always on the run," remembered Doc Bikitsha, a journalist who first met Cole in the 1950s. "He feared no one except God!" Another colleague described him as a "ninja" who managed to make himself invisible, capturing moments that no white photographer could. Much later, Cole told an American journalist that working for *Drum* opened his eyes to the world outside South Africa and brought him into conversation with other concerned Africans. Upon learning of the United Nations and the Afro-Asian bloc's opposition to apartheid South Africa, said Cole, "I decided I could help the bloc and the outside world by photographing the conditions of South Africa, the everyday life of people there."[14]

The pass system heavily constrained his mobility. His camera equipment made him particularly vulnerable to arrest: some policemen assumed he'd stolen it, while others simply didn't want their actions documented. He decided to pursue racial reclassification—something made possible, ironically enough, by the apartheid bureaucracy's obsession with racial precision.[15] To his mother's consternation, he "stretched" his hair so that he could present himself as "Coloured." He learned about the traps that bureaucrats posed to catch reclassification aspirants—trick questions like "How tall were you when you were eight?" Evidently whites and Coloureds would indicate their childhood height with palms down, while Blacks did so with palms up. Cole loved telling how, during the final interview, he teased the bureaucrat by hemming and hawing, finally drawing himself up to his full stature and placing his hand at chest level, palm down. The hapless bureaucrat caved, and the photographer walked out with a new race, a new identity card, and a more English-looking spelling of his surname.[16]

Over the course of several years, Cole managed to enter nearly a dozen compounds, often gaining entry by befriending a Black guard. "Sometimes," he wrote, "I showed up so often the guards assumed I worked there. Once in, I was rarely interfered with. To the white guards, as to the mine official, I was just another K— and they paid no attention to me."[17] He couldn't gain access to the shafts without a job ticket, so he stayed on the surface. Openly carrying his camera would have given him away. Instead, he smuggled it under a stack of sandwiches in a paper bag, which he'd outfitted with a flap that he could flip up to take a quick shot. This enabled him to document the lives of Black migrants on the compounds from the moment of their arrival to the day of their departure. Writing about Cole's photographic practice, Sally Gaule comments that his images demonstrated "speed and dexterity with the camera coupled with an alert sense of timing."[18]

Cole's images showed crowded lines of men patiently shuffling through the humiliating rituals that employers imposed in the name of profitable efficiency. Mine workers in Cole's images spent much of their aboveground time waiting. Waiting to be fingerprinted. Waiting for an assignment. Waiting for transport. Waiting—naked, facing a wall with raised arms—for a medical examination. Waiting, twice a day, to have mish mash porridge literally shoveled onto their tin plates. ("Each man must show a job ticket," Cole explained to his American readers; "only those who have worked may eat.")[19] Waiting for access to a tap to wash their clothes. Waiting to go home. And later—much later—waiting for care for the injuries and diseases they'd incurred. Apartheid and its economic engines both created and depended on a politics of waiting, in which "the powerless wait [for] the powerful [to] have time for them."[20] As an instrument of governance, waiting performed daily residuality, dividing people into those whose time was valuable, and those whose time—and personhood—could be wasted.[21]

Cole's description of how mines treated injured workers encapsulated how residual governance functioned as an instrument of racial capitalism:

> The man is returned to the WNLA depot for hospitalization, cancellation of his contract, and discharge. The mine company arranges for him to get compensation, sometimes in installments, which presumably protects the improvident African from spending all his money at one time or in one place. But it also means that outstanding balances need not be paid over on a man's death and that, even in incapacity, a man does not have freedom to go his own way.
>
> The scale of payment is frugal. For losing two legs above the knees, and thus his livelihood, one fellow I saw received $1,036, which was being paid out at the rate of $8.40 a month and was supposed to last him the rest of his life.[22]

Cole concluded his mine series with a vista of the Johannesburg skyline in the background and an untended mine dump in the foreground: waiting and dust as the twinned wastes of apartheid modernity.

Cole's laboring, waiting men were more likely to see the dumps as monumental mausoleums than as majestic mountains. Black South African writer Peter Abrahams, best known for his 1946 novel *Mine Boy*, had captured such a vision in his 1938 poem "Fancies Idle," in which he compared the dumps to ancient tombs (also built under appalling labor conditions).[23] Even before they became agents of affliction, these pyramids spoke lungs. For most who gazed upon them, they materialized grief and anguish.

The dumps carried different emotional valences for white South African artists and writers—even for the progressives among them. They shimmered as eerie geometric specters behind miners' cottages and abandoned headgear in David Goldblatt's famous black-and-white images. Cole had to choose exile to publish his photos. Goldblatt, however, was protected by his white skin, notwithstanding his opposition to apartheid and how Judaism diminished his whiteness. His collection *On the Mines* appeared with a Cape Town publisher in 1973. A first version had appeared in *Optima*, a "liberal" magazine sponsored by Anglo American "in the interests of mining, industrial, scientific and economic progress." Then-future Nobel Prize winner Nadine Gordimer, also white and Jewish, coauthored the *Optima* piece. "While certainly not serving as corporate PR," historian Alex Lichtenstein comments, "at its origins the Goldblatt-Gordimer collaboration had the blessings

3.2-3.5 Ernest Cole, from the series *The Mines* in *House of Bondage*, 1967. These images couldn't be shown in South Africa when they were taken. Today, the Apartheid Museum in Johannesburg has a long gallery dedicated to *House of Bondage*, including large-format prints of several of these images. Ernest Cole/Magnum Photos.

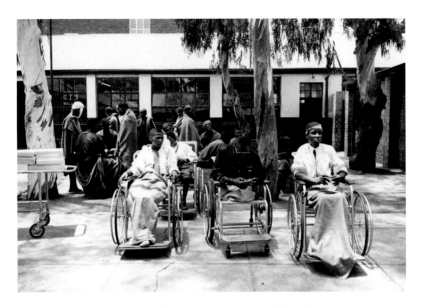

3.6 Ernest Cole, *House of Bondage* series, 1967 (not published in the original book). Ernest Cole/Magnum Photos.

of the mining house, if only as an apparent liberal commitment to free speech in the darkest days of apartheid censorship."[24] These blessings enabled the pair to publish at home, an option closed to Cole.

Like Peter Abrahams before her, Gordimer found the pyramid metaphor irresistible. In the preface to the 1973 edition of *On the Mines*, she wrote, "There was even a smell to it all, a subterranean pollen-scent of chemicals, as of the minerals flowering underground. The forms were as austere as Egypt's; but these pyramids of tailings entombed no lost civilization. It was ugly. . . . But sometimes it became perversely, suddenly, the parody of picture-postcard beauty. The dust put a red filter over the suspended sun; the step-pyramids and cones were repeated, upside down, in the lakes of dead water. Where the water was shallow it shone mother-of-pearl in its impurity or left a brilliant verdigris on the sand."[25] She wrote, too, of the Black men who migrated to the Rand for work, of white communities who lived in company towns, and of the racial inequalities in pay, working conditions, and living quarters. She imagined the plateau "long ago," before human habitation, when "white-tailed Gnu, Blesbok, Springbok, Hartebeest, Eland and Quagga roamed there." This African Eden changed when, "lugged and rocked across seas and veld from Europe . . . the machine age was unloaded on a place that had missed it out, like so much else."[26]

The distorted metrics of apartheid made Gordimer seem radical in 1973. Four decades later, her "second thoughts" for the book's new edition took stock of what she had failed to grasp earlier: "I had been conditioned by being white to see . . . 'the blankness of the place on earth,' 'a place without a past' where the mining entrepreneurs, 'the strangers found themselves.' The place where the indigenous inhabitants, black, African, had lived their own tribal social order, own activities, ancient cultures for centuries. . . . The tailings: a racially divided inhuman society."[27] Reckoning with tailings, Gordimer came to realize, meant reckoning not just with the time of gnus, not just with the time of modernity, but also with the time of her own life and her place in the mined, racialized orders that structured the lives and landscapes of the continent's southern tip.

Tailings have a distinctive temporality. The piles may look like mountains or pyramids. But they erode much faster, an erosion measurable in days, months, and years, rather than millennia: an erosion that can be witnessed live, in human time. Left to their own devices, the particles set in motion by wind do not discriminate. Dust clouds have plagued the Rand for as long as there have been sand dumps; their noxious infiltrations have affected white residents as well as Black ones, including in recent times. "There is nothing you can do when that wind starts blowing. . . . [The dust] contaminates absolutely everything."[28]

Nevertheless, the spatial character of residual governance turned dust clouds into agents of discrimination—before, during, and after apartheid. As the agglomeration took shape over the course of the twentieth century, new white communities tended to sprout upwind of the dumps, while Black, Asian, and Coloured communities were placed downwind. After 1994, new public housing was sited on available land, often immediately adjacent to tailings dams. The need for shelter perpetually exceeded the available housing stock. Informal settlements also sprang up near, and sometimes on, tailings piles. Shack residents in a perpetual state of food insecurity planted vegetables in soil contaminated by dust. Crops readily absorbed heavy metals and metalloids. Schools also planted vegetable gardens to address the hunger of their pupils. A 2012 study of one such garden, located just 500 meters

3.7 (following) INTERFOTO/Alamy Stock Photo.

FANCIES IDLE

PETER ABRAHAMS, 1938

Mine dumps of the Rand
These pyramids speak hands
Torn and bleeding,
Black, hard, rocky,
Like the black earth, wind-swept and touched by time
To leave
Torn nails and twisted thumbs
And missing spaces where the first and third fingers lived.

These pyramids speak eyes
Turned dim by gas and semi-blindness by day,
Deep, thousands of feet deep in the heart of the ocean-like earth,
Then daylight
And the hardness of the sun to turn them dim.

These pyramids speak lungs,
Tortured and touched with the coat of death,
Daily piling up layer upon layer
And wrecking the soul in a lung tearing cough,
Hour by hour with the passing of the night.

These pyramids speak bitterness
Of black men,
Thousands of black men, wrenched from their mother-earth,
And turned to gold-makers for the wealth of the earth
That grant them not the right of human thought.

These pyramids
Scattered over the body of the Rand,
Mighty in their grandeur and aloofness,
Monuments of the Twentieth Century Pharaohs,
Speak the world,
Not thousands of black men,
But millions of toilers,
Welded into a rock of firm aloofness,
Like them, made of the soul of suffering;
These pyramids speak of revolt.

3.8 The Jerusalem informal settlement abutting a mine dump on the East Rand, October 28, 2014. AP Photo/Themba Hadebe.

from a tailings dam, found high levels of arsenic, lead, and mercury in the vegetables.[29]

Chapter 4 tells the story of how these dynamics played out in one community. But I'm not done with the dumps. Let's look at how their governance was made residual, despite attempts to make it central.

Walk Away

By the late nineteenth century, swirling, airborne sand had become a nuisance for white settlers. One obvious solution: vegetation to stabilize the dumps. In 1894, the botanist in charge of park planning for the growing metropolis of Johannesburg requested seed samples from London's Kew Gardens. He hoped to find plants that would "grow on, and bind together the sand, or tailings heaps, which are accumulating so fast along the Main Reef." Describing the scene for Kew's director, he explained that extracted rocks were "first treated with mercury, and thereafter with cyanide potassium." This meant that "strongly poisonous" cyanide remained in the heaps. "Blown about by the strong winds here, the sands cause serious eye complaints, and illness. The question is, will any vegetation grow on such poisonous mountains—for so the tailings heaps may well be called."[30] Clearly, you didn't need twentieth-century science to know that the mountains were toxic.

By 1911, fifty-two mines formed a band nearly 100 kilometers long from Randfontein to Springs. Their dumps continued to grow, and to become more dangerous. The introduction of cyanide leaching into the recovery process required milling the ore more finely. Smaller dust particles were more mobile; transported by the wind, they readily infiltrated cracks and crevices. Communities complained. To damp down the dust, new regulations required mines to spray the dumps. The industry experimented with spraying a variety of substances, including molasses, salt water, clay, and night soil. The latter seemed particularly promising: trap one sort of waste with another! But the tailings dams expanded too rapidly to keep up with, and there was only so much shit to go around. The other dampening candidates didn't scale up easily either.

Without much enthusiasm, mine operators continued to test different means of covering and stabilizing the dumps, searching for species that could survive the acidic medium. Woody plants fared better than grasses. Eucalyptus seemed especially promising. Brought to South Africa from Australia in the early nineteenth century, the trees grew quickly and vigorously, making them a good source of timber to shore up shafts and stopes. Foresters hated them because they accelerated soil erosion. But eucalyptus had other useful properties, including the ability to recover from insect, fire, or frost damage.

In the absence of pollution regulation, however, research into tree-based dump covers faded. So much so that when complaints about dust from the dumps ramped up again in the 1930s, one of the mining houses enlisted a grassland botanist rather than a forester to study options. Institutionalized forgetting and expert parochialism meant that his experiments only considered grasses, despite the earlier evidence that woody species fared better.[31]

The passage of the 1956 Water Act caused consternation among mine magnates.[32] This was the first piece of legislation to hold industrial water users responsible for their pollution. Rumor had it that air pollution regulation—including measures targeting dump dust—would come next. In the hope of deflecting further regulatory zeal, the industry began looking into preemptive dust control. Stabilizing mine dumps and slimes dams would provide visible evidence of their good intentions. In the late 1950s, the Chamber of Mines created a Vegetation Unit, led by white horticulturalist William Cook. His mission: to find low-cost

3.9 Cultivating by handheld hoe and rotary hoe. Archival images from the now-closed Chamber of Mines archives. Reprinted from Reichardt, "The Wasted Years."

means of "consolidating the surfaces of mine dumps, with a view to reducing the run-off of polluted rainwater and eliminating, if possible, the blowing of dust."[33]

Cook immediately got to work. He identified a few grass species that could survive on the flat tops of slimes dams after the application of agricultural lime (to raise pH) and fertilizer (to provide nutrients). Sand dumps posed greater challenges. Among other things, windbreaks had to precede any vegetation, which otherwise would get uprooted by winter winds. Cook identified some candidate species, but cautioned that his findings remained preliminary, and that he hadn't yet found good solutions for the slopes.

Nevertheless, mining houses seized on Cook's preliminary results. Dust complaints were on the rise. The town engineer for Springs, in the East Rand, declared mine dust a "national problem." The Chamber's Gold Producers Committee worriedly reported that "authorities say that today there are more cities and towns in South Africa that are affected by dust from slimes dams and mine dumps than there are towns with a smoke problem. They think that if nothing is done to combat mine dust, it will within 40 to 50 years become impossible to live in those areas."[34] Anxious to limit the scope of regulation and prevent cessation-of-work orders, dump owners eagerly requested Cook's expertise, as well as the grassing services of his mobile unit.

Grassing the dams required labor, tools, and time. It also required exceptions to apartheid's spatial systems. After all, Cook didn't till the dumps himself. That wasn't a job for any scientist, let alone a white male scientist in the Union of South Africa. Cook explained to the Department of Bantu Administration and Development that it was "desirable, and in fact necessary, for the staff of the unit to live in camp at the sites on which the units are working. This has given rise to difficulty, since not only are the Bantu staff of the unit not necessarily registered in the areas in which they work, but your Department's inspectors must take note of the fact that they are encamped and not in compound accommodation."[35] It took months of back-and-forth to finally secure permission. Extant archives do not document the experience of the Black workers employed to grass the dumps. Historians are left to imagine the cold, dusty, backbreaking labor from the clothes and postures of the men in the photos—photos whose primary purpose was to illustrate machines.

In his detailed history of the Vegetation Unit, Markus Reichardt argues that the Chamber's move to service provision came much too soon. As Cook himself kept reminding anyone who would listen, all he'd managed to do was to establish plantings. Most did not last for more than a year or two. Fluctuations in the acidity of the dumps made the perdurance of plants particularly challenging. Cook, moreover, was not a soil scientist. The Vegetation Unit badly needed a pedologist to research plant survival. But the Chamber refused to increase the unit's budget. In Reichardt's devastating judgment, none of the men overseeing the unit "was intellectually equipped to even ask the right questions about the technical aspects of the Unit's work."[36] They wanted a low-cost solution that would free the mines to "walk away" from the dumps. Government regulators had accepted the Chamber's argument that the Vegetation Unit's approach represented the "best practicable means of dust pollution control."[37] So why do more?

Minimalism remained the holy grail of residual governance. But it was elusive. The Vegetation Unit had grassed almost half of the 8,500 hectares of abandoned tailings by the mid-1970s. But these efforts did not offer the cheap walk-away solution that companies craved. Plantings had to be carefully tended in order to stay alive and keep dust down. Would industry assume liability for their dumps in perpetuity? What could *perpetuity* possibly mean in the temporal clash between fast capitalism and slow violence, between electoral cycles and Anthropocenic acceleration?

Meanwhile, the consequences of inadequate containment continued to accrue across the industry. In 1974, a particularly heavy rainstorm burst a tailings dam at the Bafokeng South platinum mine, releasing 3 million cubic meters of toxic slurry. The slimes flooded a mine shaft, killing twelve workers underground. They rushed over adjacent land like a tsunami, submerging everything in their path. Four kilometers away from the dam, the mudflow ran 800 meters wide and 10 meters deep. Two million cubic meters of slurry reached the Vaalkop Dam, a full 45 kilometers away. Investigations concluded that "piping"—a form of erosion that occurs where layers of coarse and fine material meet within the dam—was the underlying mechanical cause of the failure. But because the burst had followed heavy rain, the inquest declared the disaster an act of God and released the operator, Impala Platinum, from liability.[38] In this and other cases, courts helped to enforce the residuality of governance.

Nevertheless, the close financial call made industry ever more desperate for a permanent, walk-away solution. In the late 1970s the Chamber latched on to promises of an untested technique propounded by an engineer who claimed that he could seal off the dumps permanently. A decade of tinkering did not deliver the promised solution. Meanwhile, the Chamber let its vegetation expertise atrophy. Other players got into the grassing game. After a few years of uneasy competition, the Chamber spun its Vegetation Unit off into the Environmental Mining Processing Rehabilitation Consultancy (their motto: "We Serve the Mining Industry").[39] The Chamber, Reichardt observes, "never again regained the initiative to position the industry practice ahead of legislative requirement."[40] He sees in this history an enormous, wasted opportunity. While Cook's work had represented the cutting edge of international mine rehabilitation research in the mid-1960s, by the early 1990s South Africa had fallen far behind.

Pollution continued to accumulate, sometimes slowly, sometimes explosively. Two months before South Africa's first democratic elections in 1994, the most spectacular dam failure yet devastated the mostly white working-class town of Merriespruit, in the Free State Province. A 31-meter-high slimes dam collapsed at the Harmony Gold mine in February. Over 2.5 million tons of slurry ripped through the nearby town, killing seventeen people, injuring another six hundred residents, and destroying or seriously damaging nearly three hundred houses.[41]

The judge who headed the inquest declared the dam "a time bomb waiting to explode." A 1985 study had determined the dam sat on unstable ground. Following steep drops in the price of gold, along with higher wages thanks to the success of a massive strike in 1987, mines across the board tried to make up for lost profits by spending even less than usual on tailings deposition. In 1991, signs of erosion began to appear in the Harmony dump. Inspections flagged drainage problems. Harmony had agreed not to deposit any new tailings on it. But "either the order to stop deposition . . . was not properly communicated to all levels of staff, or some simply ignored the order and rogue deposition of tailings took place without anyone realizing the danger of these actions."[42] (Blame rogue operators!) In any case, the entire dam failed to meet the regulatory directive to build for a hundred-year storm. Among other violations, "the pond was partially filled with metallurgical plant water, which should not have been deposited on the dam." The judge accused witnesses of painting a "distorted and false picture of events" and telling "shameful lies." He pronounced six senior officials guilty of culpable homicide through gross negligence. The two companies in charge of the site established a 10-million-rand fund to address claims. But they refused to admit liability, insisting that this was merely a "humanitarian measure."[43]

Wage gains come at the direct expense of pollution control. People can be sacrificed; profits cannot. It takes a spectacular disaster like Merriespruit to reveal (just for a moment) the inner workings of how the proverbial opposition between jobs and environment—an opposition typically presented as inescapable—is manufactured. Far more often, the details of its fabrication remain hidden in money unspent and satisfied shareholders.

Buffering the Disamenities

To mitigate profit losses caused by their own pollution, mine houses turned to the land they occupied. The more Joburg and its townships expanded, the higher the potential value of that land—provided, of course, that the land could be used for something other than harboring contamination. In her study of Crown Mines, the mining belt's largest landowner, geographer Siân Butcher describes the complex arrangement of expert knowledge and changing property rights that shaped land use in Johannesburg over the twentieth century. Starting as a fusion of many small companies, Crown Mines grew into the world's most profitable gold producer, a position it held for fifty years. Butcher notes that "Crown's mining infrastructure and waste—dumps, slimes, stopes, ore-passes, headgear, timber plantations, labour compounds, company housing—defined the valley south of Johannesburg for many decades."[44]

Crown held 13 percent of Johannesburg's land. It also, Butcher argues, held the "territorial knowledge" required to manage land use, especially all-important geotechnical data on the underground. City government and civil society organizations could readily map the surface. But two-dimensional maps didn't suffice for decision making: if too many tunnels and shafts underlay a piece of terrain—or if these hollows lay too close to the surface—building on that land would lead to subsidence and collapse. Crown's geotechnical maps of the hollow Rand were critical for the stability of surface construction.

When gold prices dropped, land development offered a path to ongoing profitability. In the 1960s, Crown spun off its property management into a subsidiary called Rand Mines Properties (RMP), charged with determining—and executing—the most profitable use of nearly 14,000 acres' worth of land. Even when the state expropriated portions of this land for motorways, RMP's "territorial power" enabled it to shape road infrastructure in its favor, steering major throughways away from white amenities such as golf courses.[45]

Mines had long figured as decisive elements in the racialized infrastructures of the ever-expanding metropolis. Patterns repeated up, down, and across spatial scales, deepening through each new iteration. White residents in the northern (upwind) suburbs tempered the visually harsh Highveld with British-style manicured lawns that performed the privilege of whiteness.[46] Planners of Black (downwind) townships did sometimes incorporate green elements into their designs. Architect Hannah le Roux uncovered early plans of KwaThema that include garden plots: productive planting rather than decorative vegetation, but green all the same.[47] As with so many well-laid plans, however, the reality didn't match the ideal—particularly in the early years of formal apartheid, with government whites riding high on the victory of their violent vision.

Cycles of evictions, forced displacements, and eternally inadequate housing took their toll. Squatter movements, such as the one led by James Sofazonke Mpanza in the 1940s, filled some of the gaps with informal settlements. In the absence of land tenancy and state permission, however, these living arrangements were inherently precarious.[48] The demand for housing continued to outstrip supply.

Segregation was already embedded in urban infrastructure when the National Party rose to power in 1948. Grand apartheid entrenched it further by using buffer zones to maximize the distance between European and Native homes while maintaining access to Black labor. Outlining his plan to Parliament in 1952, Minister of Native Affairs Hendrik Verwoerd mercilessly prescribed "the provision of suitable open buffer spaces

3.10 Forced removal under apartheid. *House of Bondage*, 1967. Ernest Cole. Ernest Cole/Magnum Photos.

around the proclaimed location area, the breadth of which should depend on whether the border touches on a densely or sparsely occupied white area, and a considerable distance from main, and more particularly national roads, the use of which as local transport routes should be discouraged."[49] White anxiety about the "hygiene" of Black Africans had propelled forced removals since the early twentieth century. Apartheid rule simply made instruments such as US redlining practices unnecessary. Cruder tools inspired by imperialism (such as the *cordons sanitaires* Europeans used to redesign colonial urban spaces) would work just fine to spatialize racism. Segregation laws piled on fast and thick, restricting where Black Africans could live and how they could move through town. By their very nature and composition, mines and their dumps served as buffer zones. Their sheer spatial extent amplified the brutal racialization of urban space.

Mining houses shaped urban geography more directly, too. In 1956, for example, magnate Ernest Oppenheimer poured £3 million into "slum clearance," openly financing the forced removal of Black residents from their homes in spaces newly declared white.[50] Architects and town planners, meanwhile, reinforced the strategy of using mine lands as "buffer zones." Consider a proposal commissioned by Crown/

RMP in 1969 to develop Ormonde, a new suburb for upper-middle-class whites that some feared came too close to Soweto. The proposal sought to reassure white town planners:

> The proximity of the site to Soweto cannot be overlooked. The existence of Soweto poses a number of environmental problems, these need not be detailed, but provisions to counter them should be outlined. Basically, all such provisions can be made in combination by the provision of a suitable hierarchy of buffer zones between the site and Soweto. A logical pattern of buffer zones would consist of an industrial zone bordering Soweto with due precautions to protect the residents from the disamenities [*sic*] of the industry, a band of open space, the Western bypass and finally, on the site, a judiciously designed screen of trees, shrubs, and other landscape features such as mounds and areas of open space.[51]

These fantasies took successful dump vegetation as a given. Evaluating this proposal four decades later, architect Jennifer Beningfield highlights its treatment of landscaping as "primarily a white amenity," accentuated in the accompanying plans by greenlining: literally coloring zones around and within the proposed community green. She further observes that the very viability of the plan depended not only on racial disparity in the present, but also on its imagined perpetuity. Offering projections of nationwide population growth, the plan predicted that by 2000, three-fourths of all Black people would have "moved or been moved to the Bantu homelands."[52] The passive (dispassionate!) voice conveniently elided the violence of forced displacement.

Casting Soweto as posing "a number of environmental problems" that "need not be detailed" casually betrayed the callous arrogance of apartheid planners. Cemented into buildings and roads that were difficult if not impossible to dislodge, racist buffers would have a long life. Half a century later, a 2018 report by the Bench Marks Foundation tellingly titled "'Waiting to Inhale'" spelled out how urban planning's ongoing deployment of mine spaces (and materials) embedded contamination in the lives of residents:

> Historically, those residing in Soweto and Riverlea . . . had no choice about where they wanted to live; they were forcibly relocated from other areas that were safer, with healthier environmental conditions than Soweto. The residents also had no choice in the size of yards, design or building materials used in the construction of their houses. Thus, they had no choice in the roofing materials used (asbestos). They also had no

choice in the location of their houses or the distance of their homes from toxic, radioactive mine dumps. Nor did anyone ever inform or educate them about the implicit dangers associated with operating abandoned, derelict, and ownerless mines.[53]

Like the acid water bubbling up from the shafts, the slow violence of tailings dust has become infrastructural, another piece of code in the algorithm of the new apartheid.

Residual Value

As long as the gold price remained low, grassing the dumps and turning mine lands into residential or industrial spaces made financial sense. But changing times could shift the status of some dump components from waste to resource. That's what happened soon after atomic bombs over Hiroshima and Nagasaki marked the end of World War II and set the technological terms of the Cold War. In the late 1930s, Peter Abrahams (along with hundreds of thousands of workers) had seen the residual pyramids as sepulchers of suffering. A decade later, mining magnates contemplating the colossal remnants of gold gluttony saw vaults of value: gigantic piles of cheap uranium.

The scramble for uranium rode the tails of wartime urgency. Geologists had not previously paid much attention to the element, so initially they believed it was rare. The US fantasized that it could secure a worldwide monopoly on the element if it moved quickly enough. Geological data suggested that South Africa might possess the largest uranium reserve in the world, with a great deal already extracted and sitting in the mine dumps, and still more underground. In 1950, the US and the UK contracted to buy 10,000 tons of South African uranium. The terms of the agreement offered South Africa excellent prices for ore and substantial loans for infrastructure development. For the mining industry, the timing was impeccable: the price of gold had tanked, and some companies were facing huge losses, even bankruptcy. Without the uranium contracts, many mines would have folded.[54]

The significance of uranium extended well beyond this economic bonanza. In 1948, Prime Minister Jan Smuts addressed Parliament at great length, extensively detailing the technical history of South African ore extraction to argue for an inevitable logic to his nation's participation in "the biggest scientific discovery that has been made probably for

hundreds of years."[55] National Party leaders eagerly adopted this narrative when they took power later that year. In 1952, West Rand Consolidated Mines opened the country's first uranium-producing plant in Krugersdorp with considerable pomp and circumstance. Apartheid's first prime minister, Daniel Malan, hailed South Africa's determination "to make our contribution to the cause of the Western Powers."[56] The plant's celebratory brochure—peppered with pictures of white men in suits and lab coats—proclaimed its opening to be "the most important metallurgical event of the century," one that would shape "the future destinies of South Africa."[57] The fulsome declamations erased the participation of Black workers, offering a surfeit of technical detail on plant design but none on its construction or operation. More factories followed. Over time, apartheid South Africa would also sell uranium to France, Israel, Iran, and Japan.

Some of the dumps disappeared into the uranium scramble. But not all pyramids contained enough of the radioactive rock to be worth remining. When the price of gold began to climb in the mid-1970s, the industry eyed the remaining piles of residue as fresh sources of gold ore. New chemical and metallurgical techniques made it possible to mine the dumps as low-grade surface deposits. Extraction of their gold once again promised profit.[58]

At least one individual had anticipated this moment: a Portuguese immigrant named Jose Manuel Rodrigues Berardo. As he explained to a journalist in 1980, "I used to see those big 'mountains' and I always thought the gold price would go up. . . . I began to apply for the rights on some dumps in 1973. A lot of people wanted to get rid of them. I bought some very cheap. 'I will try to move this mountain,' I told them. At the time they thought it was impossible. They thought I was a bit cuckoo. But now everybody is fighting for what I've got."[59] Berardo's vision ran from the peculiar to the prophetic. He foresaw a return to gold as currency. "How long can the governments go on printing paper money that has no value?" he wondered. "I don't think it will work out in the long run." Even if that didn't come to pass in his lifetime, though, he had other ideas for how to turn a profit. "I think I'll put sand from the dump, after I've treated it, into little bags that say, 'This contains .0002 percent of gold' and sell them in America. If Americans will buy cans of fresh air, and I've seen them do it, why won't they buy bags of gold? I think they will." The man had a point, even if he lacked a feasible marketing and distribution plan.

Mining houses had far more ambitious plans for how to make money from the dumps. Anglo American led the charge. In August 1976—just six weeks after the slaughter of Soweto's children grabbed global headlines—Anglo launched a scheme to extract gold, uranium, and pyrite from mine dumps in the Free State. In 1978 it built a new treatment plant, spun off as the ERGO project, designed to reprocess material from sixteen dumps across the Rand. Compared to the cost of sinking new shafts, the capital invested in the new plant seemed low. Operations quickly recouped their costs.[60] According to one estimate, the South African government collected some $2 billion in taxes and reserve bank gold sales in 1979, three times what it had anticipated. The bonanza made international news: in 1980, Anglo's chairman "exulted" to the *Washington Post* that one mine had produced 80 kilograms of gold just by cleaning up its railroad line. "Some people say it's time to tear up the streets of Johannesburg," he joked, "but I'm not in favor of that."[61] By the time the ERGO plant paused operations in 2004, it had processed 890 million tons of tailings, yielding 8.2 million ounces of gold and 5.5 million pounds of uranium.[62] (Operations restarted in 2007; I'll get to those later.)

Unease generated by world events in 1979 (the Iranian Revolution, the Soviet invasion of Afghanistan) continued to drive gold prices higher. Rand Mine Properties jumped into reprocessing, building a retreatment plant at Crown Mines in 1980 and others later that decade. But it didn't abandon the real estate game. Throughout the 1980s, RMP continued to establish industrial townships, shrewdly lavishing municipal officials and developers with hunting expeditions and golf games. The resulting settlements, Butcher notes, "consolidated apartheid buffer zones" while keeping territorial knowledge firmly in RMP hands.[63]

With ups and downs, tailings reprocessing continued into the twenty-first century. South Africa's weakening rand increased the local value of exports such as gold. Not to mention uranium, which had seen its price skyrocket from $10 per pound in 2003 to over $90 per pound in 2007. Junior producers (that's mining business parlance) jumped into the reprocessing fray. Mintails was one notable newbie. Formed in 2005 and immediately oversubscribed on the Australian Stock Exchange, the company aimed to cash in on surface tailings via a variety of ventures with Durban Roodepoort Deep (DRD) Gold. Deploying standard spinoff and subsidiary strategies, it acquired Mogale Gold on the West Rand and the old ERGO plant on the East Rand.[64] "The retreatment business is high-volume and low-risk," the company told investors, because it didn't require sinking capital into a long-term project.[65] "You can effectively achieve a financial payback in two or three years," explained another executive.[66] The short lead time from production to profit enabled mining companies to respond rapidly to price and currency fluctuations. Reprocessing was profitable for residual ore grades ranging from 0.2 to 0.4 grams of gold per ton. By comparison, underground grades had to reach at least 10 grams per ton to turn a profit. As retreatment became the main game in town, underground giants like AngloGold began withdrawing from the gold business.[67]

But others remained. Harmony spun off a new subsidiary by selling a 60 percent stake of its Randfontein uranium assets to Pamodzi Resources, a Black-owned private equity fund. This enabled Harmony to score coveted Black Economic Empowerment points, putting it in the government's good graces. To fund the deal, Pamodzi applied to the Industrial Development Corporation (IDC), the state's development financing arm, for 400 million rands. The IDC gave 200 million in an escrow account, sending Pamodzi to the market for the rest.[68] Over a billion dollars' worth of funds from US investors enabled the company to boast of its "proven ability to attract long-term foreign direct investments to South Africa."[69] In reporting the story, *Business Day* joyfully declared, "The West Rand will get a major boost for jobs and environmental rehabilitation as Harmony Gold Mining has reached a decision on treating the slimes dams in the area to extract their uranium."[70] Black empowerment, jobs, and environmental cleanup: What could possibly go wrong?

In this case, everything. Scarcely two months after the IDC's escrow deposit, a Pamodzi employee took the money and ran. The company went bankrupt. Lumkile Mondi, the IDC's chief economist at the time, notes ruefully that the mess on the Rand worsened rather than improved after 1994 because of such shenanigans.[71] State financing could be justified if environmental cleanup accompanied the process of remining and retreatment. All too often, however, mine executives played complex shell games that enabled them to strip the dumps of assets, then escape before shouldering the cleanup. Almost as often, the state was unable to hold them accountable. Or unwilling. I could spend the rest of this book narrating the notorious networks that joined Jacob Zuma (the nation's deputy president from 1999 to 2005, and its president from 2009 to 2018) to smarmy asset strippers like Brett Kebble or the ultra-wealthy Gupta family. South Africans describe these sorts of relationships as "state capture."[72] By this, they mean that oligarchs have captured the state—not just its regulatory agencies, as conveyed by the

3.12 Formally employed workers remining Mennell's Dump, southeast Johannesburg, 2007. The workers are surrounded by dust and sand, but there's no evidence that they were provided with respiratory equipment. AP Photo/Denis Farrell.

American term *regulatory capture*—but most or all of the executive branch. Given the wealthy world's tendency to portray the continent as inveterately corrupt, let's pause to note that oligarch networks are by no means uniquely African. State capture in the South African sense can apply on any continent, including North America.

There's no doubt that state capture has contributed mightily to the super wickedness of the Rand's residue problems. I'll return to this at the end of the book. But reducing the analysis to such an explanation would be a serious mistake. Historically, politically, and materially, the problems of residual governance far exceed the corruption of power-hungry men and resource-starved bureaucrats.

Whose Heritage?

One tailings pile targeted for reclamation was not like the others: the Top Star dump, perched on the southern edge of Johannesburg's central business district. Originally known as the Ferreira dump after the mine that fed it, the pile started in 1899 and accumulated over the course of four decades, eventually covering nearly 43 hectares.[73] A range of discarded materials—including construction waste and household trash—served to cap the dump. After the extraction operation shut down in 1946, the property owners received permission to change the zoning from "industrial" (allowing "many thousands" of Black workers access to the area) to "general business" (limiting the "influx" of Black workers to two or three hundred).[74] Rezoning supported the Johannesburg town council's plan to expand the business district southward, pushing Black residents even farther out and widening the waste buffer.

The council balked when the owner applied for permission to develop a township on the property in 1950. Concerned about potential land subsidence, and especially about the council's liability in the event of such a disaster, Labour Party councillors called for a commission "to investigate the whole matter of building on old mine dumps."[75] In response to this and a range of pollution complaints, the owners leveled the top and planted trees and flower beds.[76] After nearly five years of discussion and negotiation, the council permitted the township in 1955, subject to an indemnity clause that released local authorities from responsibility for any consequences stemming from "the subsidence caving or sliding of the slimes, sand, rubble or debris covering the land, whether as the result of natural causes, mining operations, past or future, or any other cause whatsoever."[77] In short, the town council abdicated any governance responsibility for the dump and its afterlives.

As long as the city didn't have to pay for damage, plans could go ahead. Fortunately for potential inhabitants, this township was never built. A few years later, the city approved a new proposal to build a whites-only drive-in movie theater on the dump—or, as the permit application put it, a "Bioscope (European)." Another layer of topsoil was added to the slopes. Unlike the meager plantings that topped dumps with no commercial purpose, Top Star's vegetation was meticulously maintained, doubtless with considerable labor from Black men.[78]

With its futuristic entry gate, gigantic screen, and breathtaking city views, the Top Star drive-in quickly became a cultural fixture for "(European)" Joburgers. The apartheid government managed to keep television out of the country until 1976. So for over a decade, crowding into the family car and heading to the drive-in was the closest white households could get to the experience of watching a movie in their homes. And what a scene! One moviegoer especially remembered moments of family intimacy afforded by the drive-in: "The Top Star for me meant

3.13 Front gate of Top Star Drive-In, 1994. François Swanepoel. https://www
.flickr.com/photos/61506334@N04/24972662838/in/photostream/.

Friday nights when I had my mom and dad to myself. I always remember the gold of the city lights on the highway making bands on my arms, the smell of my blanket and comfy pillows in the back of the old Alfa. The winding road that took you to the top of the mine dump. It was so desolate that you thought you had gone astray . . . and then the Drive-in greeted you, and Johannesburg was a glittering map below."[79] Restless children played on the swings and jungle gym at the foot of the massive screen, while younger siblings fell asleep on the back seat.[80] Crafty teenagers and twenty-somethings would "load a few friends in the boot" of the car to avoid entry fees.[81] Some viewers got so cozy (or inebriated?) that they forgot to return the audio speaker to its pole before pulling out, which meant that "either the speaker was ripped out of the pole or the side window was shattered."[82]

In 2006, the new owner of the dump, DRD Gold, announced its intention to close the theater and "reclaim as much gold as possible" from the pile. The news caused consternation among those who'd enjoyed watching the city unfurl as they wound their way up to the parking lot in the sky. Artists memorialized Top Star's iconic status in painting, film, photography, and literature. Some expressed painful awareness of nostalgia's contradictions. For writer Mark Gevisser, the mine dumps "were the mountains of our childhood, covered in grass and planted with trees." Only as an adult did he understand their role as "an obvious buffer between the white parts of the city and the Black ones." Only then did he see his childhood "obliviousness . . . as precisely the consequence of the type of blinkering we endured as white suburban children in apartheid South Africa. We lived in an artificial world, our own void of sorts, dug out of the earth by the hunger for gold."[83]

As part of its permit application, DRD commissioned a "heritage scoping exercise" to assess the likelihood of obtaining a permit to flatten the movie mountain. Heritage was a highly contested mode of politics in the newly democratic nation.[84] For some, heritage making served a therapeutic purpose, enabling Black citizens to reclaim their history on their terms.[85] But in a "new South Africa" consumed with the Truth and Reconciliation process, heritage as therapy immediately became entangled with heritage as politics.

Heritage making required reclaiming histories made residual by colonialism and apartheid. What counted as heritage? Which events and places merited memorialization? Whose heritage had value? How would that value be established and expressed? Who would pay for museumification? Who would benefit from the commodification of culture? All these questions carried political stakes. The National Heritage Resources Act of 1999 broadly delineated potential heritage targets and established a state apparatus to work out the details. Historian Sipokazi Madida argues that this apparent bureaucratic order hid the discrepancies, confusion, and contestation of heritage making.[86] The terrain was ripe for consultants who could steer developers through the system. Matakoma Heritage Consultants was contracted by DRD Gold to assess Top Star's heritage value.

Legislation did include mine dumps as potential objects of protection. But the grounds for conservation were unclear. Should Gauteng's history of mining be celebrated as achievement or deplored as exploitation? Which artifacts and ruins should remain? In commemoration of what? Matakoma's (white, Afrikaans-named) archaeologists specialized in such questions. They predicted DRD would face challenges in the permitting process; the dump's use as a drive-in meant that it represented a "unique social phenomenon."[87] Could DRD argue that reclamation would contribute to the continued employment of over nine hundred people and remove the dump's "visual impact"? The consultants warned that the dump had become an integral part of Johannesburg's skyline and tourist identity and that it did already generate revenue. Based on these findings, the Provincial Heritage Resources

Authority of Gauteng (PHRAG) issued a two-year protection order to halt the dump's destruction.

The second round of DRD's arguments in favor of reclamation proved more successful. Removal would eliminate a source of pollution, rehabilitate the area to "appropriate environmental standards," and "unlock" key urban land for development. Not that such considerations had motivated the project, mind you. The company merely followed the greenwashing playbook by leveraging the growing uproar around mine pollution. The consultants cautioned that to make this argument work, "pollution from the dump must be quantified." Even then, they warned, the heritage agency might deny the permit.

But DRD wasn't willing to endure a lengthy permitting process. "The clock is ticking," said their spokesman. "We will make a small margin of profit." Heritage officials countered that their clock also ticked: dumps, villages, and mine equipment spanning a century of extraction were disappearing quickly, increasing the historical value of remainders such as Top Star. The company decided to circumvent PHRAG by approaching its trusty state partner, the Department of Minerals and Energy, for a mining permit (as opposed to a demolition permit). Success: in August 2008, DRD began ripping into the dump. Abandoning all pretense of caring about pollution, its spokesman instead invoked the increasingly urgent imperative to liberate land for urban development: "This is prime land containing an inert lump of waste." (Note how quickly they jumped from "pollution" to "inert waste.") To no avail, PHRAG called the police on DRD. During the first year of remining, Top Star sand was piped to a nearby gold reclamation plant at Crown Mines. After that, it went to the ERGO plant 30 kilometers away in Brakpan, on the East Rand.

The movie screen was the last thing to go, DRD's one concession to (mostly white) longings for heritage. By 2011, Top Star was flattened. Photographer Sally Gaule captured the demolition in an exhibit titled *Stardust*, offering gallery visitors one last opportunity to engage with the dump's distinctive aesthetic. (I based my discussion of evenings at the drive-in on visitor inscriptions in the gallery's memory book, which Gaule kindly shared with me.) She ended the show with a photo of Brakpan's brand-new 23-kilometer superdump, the current resting place of Top Star's sand.

Gaule's images invited reminiscence while challenging the nostalgia of those who considered the disappearance of Top Star a "personal loss."[88] She'd heard Peter Abrahams's voices trapped inside the pyramid,

3.14-3.17 Sally Gaule, *Stardust* series, 2012: Top Star screen just before demolition; Johannesburg Central Business District, seen from the remains of Top Star; factory and storage box gesturing toward the variety of economies at work during removal of Top Star; Brakpan superdump.

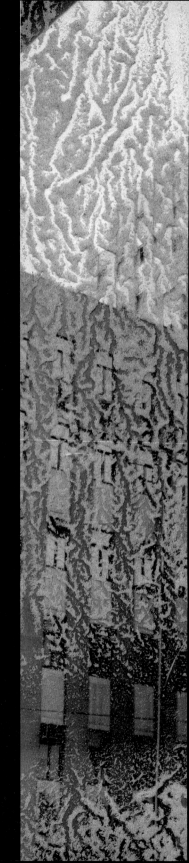

3.18 Sally Gaule, *Stardust* series, 2012: *Chamber of Mines Building with Sand*.

speaking lungs and hands; she responded by invoking the invisible lungs and hands scratching at the dying dumps nearly a century later, made visible by traces of labor that reflected "tenuous and mobile lives that contrast with the scale of surface mining."[89] Grating against the grain of nostalgia, a formal black-and-white photo of the Chamber of Mines building overlaid with sand invited viewers to imagine a world in which the Chamber, rather than (or alongside) surrounding communities, was overlain with the toxic residues of its wealth.

The Crown recovery plant lay just a few kilometers west of Top Star. Next on the Main Reef Road sat the community of Riverlea, dwarfed by the ten-story-high Mooifontein tailings dumps. Established in 1963 for Coloured families evicted from white suburbs, Riverlea was roughly two-thirds Coloured and one-third Black (plus a smattering of others). Fully half of its residents reported respiratory ailments. The zone that immediately abutted the dump—known formally as Riverlea Extension, and informally as Zombie—was particularly hard hit. Researchers from the Bench Marks Foundation found Zombie residents of all ages living on oxygen machines who "complained that their respiratory problems were worsened by the dust from the mine dumps, especially on windy days."[90] Immediately to the south, Crown Mines College was squeezed between two of the Mooifontein dumps. College staff reported a raft of chronic conditions: "Our skins are always itching, our eyes burn. We have constant bronchitis, sinusitis, headaches, and chronic fatigue. It doesn't go away." On windy days, "this whole area is coated in white." On rainy days, "we're on holiday. Then we can breathe." The lists of ailments went on and on, in community after community. A college staff member summed up the experience of many: "We were dumped here. . . . We call ourselves zombies, because that's what we feel like."[91] The pyramids definitely spoke lungs in this neighborhood.

As architect Mpethi Morojele observed, "The city's heritage means different things to different people." He remarked (perhaps with deliberate mildness) that "most Black people will have conflicting emotions about the preservation of this history."[92] Regardless of perspective, it was hard to dispute writer Mark Gevisser's elegy: "None of us would exist—the city itself would not exist—were it not for these violations against nature. There is no reason for Johannesburg before or beyond them, but now it exists, despite them. This is our inheritance."[93]

3.19 Monde, Puleng, Zizipho, and Khuselo play on the Riverlea mine dump near their homes, 2011. Ilan Godfrey.

Superdump!

Given its abysmal track record, DRD's argument that removing Top Star would have environmental and health benefits seemed opportunistic at best. Nevertheless, the extent and spread of tailings dumps across Gauteng presented considerable cumulative danger. Consolidating the tailings materials into a small number of megadumps could make hazard management more tractable. Combined with remining, the process could even generate profits. Brakpan, the resting place of Top Star's remaining sand, offered proof of concept on the East Rand.

In 2009, two mining companies took a similar plunge on the West Rand, proposing to consolidate their tailings into a giant superdump near the banks of the Vaal River. The proposed site extended over a 25-square-kilometer area between Johannesburg and Potchefstroom.[94] It received initial government approval, overriding protests from farmers and environmental activists, who objected that "the pollution is not cleaned up at the source, but passed on to other users." Mine Waste

Solutions, the company tasked with establishing the dumps, insisted that it wouldn't repeat past mistakes: "What is so good about our plan," exulted one spokeswoman, "is that all 14 existing dumps will be re-worked, levelled and rehabilitated. All the residue from these dumps will be taken to the new site for processing."[95]

Meanwhile, the company contracted to do the environmental impact assessment of the proposed dump sought to reassure the public. "World-class design" would enable builders to "re-vegetate the dump simultaneously while the dump is developing." Such statements blithely ignored the failures and lessons of previous vegetation efforts. Residents didn't buy it. They knew that "rehabilitation" didn't mean that land could be used for agriculture, or water for drinking. The Mahatammoho agricultural cooperative, established for "emerging" Black farmers, stepped up to protest the decision. Co-op spokesman Justice Lunuberg explained how the superdump would cut off chances for economic empowerment allegedly afforded by the (now not-so-new) democratic government: "Many of us as emerging farmers have just started farming and have also just received help from the Department of Agriculture with setting up boreholes on our farms. . . . When the sludge dams are set up we're going to have water pollution. Our farming vision is going to be totally disrupted. It's good if we as Black and white farmers can unite. More white farmers are now joining our co-operative. We fight with one voice."[96] At least momentarily, Black and white farmers derived common political purpose in combating residual governance along its three dimensions: physical waste, minimal management, and the experience of having their own selves, bodies, and land treated as waste.

Farmers and their neighbors spoke from bitter experience. Dust from the dumps had long spoiled their food and made it impossible to keep their homes and laundry clean. Older residents who'd inhaled dust for a lifetime had fallen gravely ill. Sixty-five-year-old Susanna Smit could barely breathe: "My doctor told me my lungs look like a man who has worked underground in the mines for thirty years . . . but I've never worked on a mine. I've never smoked a day in my life and neither has my husband." Many believed that local doctors were reluctant to link pulmonary and other illnesses to tailings dust because they feared retaliation from powerful mining companies. Residents in the most affected zones of the Rand strongly believed that the rise in cancer in their communities was linked to mine dust inhalation. As Mariette Liefferink noted, however, "The burden is now on affected communities to prove there is a link with the mining waste." But with what resources, and whose help? One farmer asked, "Must we prove this pollution with autopsies on our family?"[97]

Mine Waste Solutions obtained a license from the Department of Water Affairs in June 2011. The company insisted its proposal would benefit water quality by reducing salt deposits in the river. The clay soil under the dump would serve as a barrier, preventing leakage. Mine Waste Solutions further promised to launch a community forum. Provincial authorities gave the go-ahead. The company erected two plants to extract gold (with a combined feed capacity of 26 megatons per year) and one for uranium (1.2 megatons per year), all three fed by slurry reclaimed from other tailings facilities.[98]

The Federation for a Sustainable Environment immediately filed an appeal on behalf of communities and landowners, arguing that because the proposed dump was not lined, it would ultimately leak into the Vaal River and surrounding land. Earthlife Africa, Save the Vaal, local conservancy groups, and others also opposed the permit and complained that the company had neither consulted them nor shared any significant details about the project. The discards of diminishing dumps would continue their deposits in the lungs and imaginations of Gauteng's people. And some of those people were prepared to forge previously unthinkable political alliances to flip the residual governance script.

4

SOUTH AFRICA'S CHERNOBYL?

SOUTH AFRICA'S NATIONAL HOLIDAY is known as Freedom Day. Celebrated on April 27, it marks the anniversary of the country's first truly democratic election in 1994, when some twenty million South Africans cast their ballots, representing over 86 percent of registered voters. The vast majority were voting for the first time; people queued for hours to cast their ballots, with some lines over a kilometer long. The voting period extended over four days to ensure maximum participation. Although nearly twenty parties were on the ballot, the top three garnered over 90 percent of the votes; the African National Congress (ANC) emerged as the undisputed victor. On May 10, 1994, Nelson Mandela was inaugurated as South Africa's first Black president.

The long wait—for justice, for democracy, for equity, for resources— was finally over. Or so it seemed. Full of promise and promises, the government launched a Reconstruction and Development Programme (RDP) to tackle some of the nation's most severe infrastructural inequities. Water, electricity, health care, land reform, and public works were all targeted by the RDP. But the core of the program was subsidized housing. Over one million RDP homes went up in the first few years of the program, housing around five million South Africans. Progress to be sure, but there was still a long way to go, especially for the more than seven million who still lacked proper shelter.

Jeffrey Ramoruti was among those who had to wait. At the time of the election, he and his family were living at the West Rand Consolidated hostel. The building was targeted for demolition in 1995, forcing Ramoruti's household and some twenty-five others out. Designated as Kagiso Extension 8, the place they were told to occupy sat smack on top of uraniferous tailings and the old offices of the defunct Tudor Shaft mine. Officials reassured them that the placement was temporary. This would be their last forced relocation; the new South Africa had no place for such violence. In six months, they'd have their RDP houses. Meanwhile, they built shacks for shelter. Six months turned into a year, then two, then more. The settlement grew as new residents trickled in, reaching nearly six hundred households.[1] Ramoruti and his neighbors advocated ceaselessly for their right to proper housing and a clean environment. When I met Mr. Ramoruti in 2016, they were still waiting.

In many respects, Ramoruti's story resembles that of millions of others. Yet one key aspect stands out: over the course of two decades, the Tudor Shaft informal settlement became the technopolitical epicenter of the Rand's radioactive residue problem. It wasn't necessarily the single most contaminated place on the Rand (though it may have been— the data simply aren't there to make this determination). But it became the place that garnered the most attention for its radioactivity. One newspaper headline proclaimed Tudor Shaft and its surroundings to be "South Africa's Chernobyl," as unsuitable for life as the exclusion zone around the site of the world's most notorious nuclear accident.

At Tudor Shaft, radioactive contamination became a key flashpoint for the conflicts over volumetric violence traced in the previous two chapters—so much so that proponents of residual governance thought they could limit their measures to those that addressed radioactivity. But as Ramoruti's story shows, the conflicts were about much more than radiation. The area around Tudor Shaft posed its own wicked problem, one that enmeshed a wide range of residues and governance dilemmas.

4.1 Tudor Shaft informal settlement in 2016, with the world's highest sand dump in the background. On the small rise from which I took this photo, my Geiger counter showed radiation levels thirteen times above ICRP guidelines. While this was not an official reading, it did correspond to findings in the reports discussed in this chapter. G. Hecht.

The administrative landscape alone conveys some of this complexity. The settlement lies within the township of Kagiso, itself part of Mogale City Local Municipality, which in turn belongs to the West Rand District Municipality. This municipal nesting aims to facilitate administration of government services. But it also facilitates buck-passing, especially when combined with the large number of provincial and national agencies involved in managing mine residues. The proliferation of administrative bodies poses serious hurdles to activists and residents with limited resources.

Yet it also offers multiple entry points for advocates fighting against the treatment of people as waste. Combating their own residual status has required Tudor Shaft residents and their allies to engage in an all-fronts approach, pushing on every conceivable opening, at all possible governance scales: municipal, provincial, urban, national, and international. The struggle of Ramoruti and his neighbors offers a microcosm of the work required to overcome residual governance: not just in Gauteng, not just in South Africa, but everywhere. To apprehend this more fully, let's begin with a history of Mogale City and Kagiso.

Mogale's City

Change bubbled all around in 1994, starting with place-names and districting. Replacing the toponyms of white supremacy with those of African leaders offered a relatively straightforward way of redressing historical residuality. Mogale City Local Municipality joined the first wave of decolonial nomenclatures, absorbing the town of Krugersdorp.[2] The symbolism was clear: rightful Batswana chief triumphs, at long last, over Afrikanerdom's cherished hero. The municipality's website casts the region's history accordingly. "The young heir to the Po chiefdom of the Batswana gave his name to Mogale City, once the seat of his kingdom. Chief Mogale is believed to have ruled over a kingdom of miners and traders of gold, who had an influence in far-reaching markets. Their cross-border trade was said to extend as far north as Egypt, on the Mediterranean tip of the continent." Po rule over the region was shattered during the Mfecane. In 1822–23, Mzilikazi ka Mashobane, the chief of the Northern Khumalo clan who'd sworn allegiance to King Shaka in exchange for protection, rebelled and fled across the Drakensberg mountains, forming his own kingdom on Po land, killing Mogale's father and taking the young man captive. Rescued by his people, Mogale and his group fled south to escape Mzilikazi's brutal reign. They bided their time until Mzilikazi, in turn, was driven out by the Voortrekkers. Mogale and his people returned, only to find Boer settlers farming their land. The former rulers were forced into servitude, adding insult to already considerable injury. "Today," concludes the website, "a true son of Africa has had his name rightfully honored in Mogale City."[3]

While the RDP's infrastructural remediation plan required serious capital investment, the historical remediation offered by renaming and redistricting entailed a relatively modest financial outlay, one that fit more comfortably into stretched budgets. Such acts of heritage recovery proliferated in the newly democratic nation.[4] Indeed, Mogale City itself continued to invest in historical redress. In 2010, the municipality commissioned Vusumuzi Khumalo to write a history of Kagiso township focused on the ebb and flow of life for African residents previously deemed residual.[5]

Like other townships, Kagiso was conceived under and for apartheid rule. Built in the late 1950s, it housed Africans employed in gold mines and other facilities in the area's burgeoning industrial sector. As elsewhere, the first wave of Kagiso residents was forcibly displaced from their previous location in Munsieville. The one- and two-room family houses were built with concrete blocks and asbestos sheeting. One long-time resident told Khumalo, "Early on we got lost, and ended up knocking on the wrong door because these houses all looked the same." The houses were cramped, and Kagiso's single-sex hostels even more so. Relations between hostel and house dwellers could get tense; the latter saw the former as unruly, sometimes dangerous. But there was also solidarity. Township residents accommodated wives visiting their hostel-dwelling husbands. The two groups drank together in local shebeens and attended *isicathamiya* music competitions staged by Zulu migrant workers in the hostel hall. As in other townships, a distinctive linguistic mélange, or Ischamtho, emerged: Kagiso's version mixed isiZulu and Se-Sotho with Afrikaans. "Right from the start," Khumalo explains, "these languages fill whatever gaps speakers encountered in each vernacular."

In step with the rest of South Africa, political action in Kagiso intensified in the mid-1970s. Churches, central to many facets of community life, also served as conduits for the United Democratic Front, the Azanian People's Organisation, and other groups involved in the liberation struggle. On June 17, 1976—the second day of the student uprising that began in Soweto—students and adults demonstrating in Kagiso were forcibly pushed back by police. The crowd responded by attacking West Rand Administrative Buildings; the police summoned reinforcements and shot into the crowd, killing five people.[6]

The following year, the apartheid state created Community Councils to manage township life, hoping to co-opt Black community leaders by creating elected positions. These, in turn, opened opportunities for graft. Merciless on this point, Khumalo notes that many councillors sought office to enrich themselves, spending community funds on cars, liquor, and other frivolities. After draining the coffers, officials raised the rent on township houses. Large families spilled out of their tiny homes. "Where does big sister do her homework when the bedroom is shared with the talkative little ones, the living room choked with the beer and smoke fumes of men playing cards?"[7] Grievances grew. People stopped paying water bills. Resident organizations grew larger. People marched to protest increases in rent and bus fares. The police responded with violence. And they got it back, in kind: in 1985, Kagiso's commemoration of the Soweto uprising ended with burned buildings and stoned police vehicles.

Data from South Africa's 2011 census, collected the year after Khumalo completed his study, clarify the stakes and limits of redressing historical narrative. The population density in white-majority Krugersdorp was just under 570 people per square kilometer, while, in almost entirely Black Kagiso, it exceeded 8,170 people per square kilometer.[8] Mogale City may have subsumed Krugersdorp, but well into the twenty-first century, Black residents enjoyed far less space than their white counterparts: a stark illustration of the need for land reform. In this and other ways, Mogale City offers a metonym for the planet's predicament, propelled by the brutal reality that wealthy humans leave much larger carbon and environmental footprints than humans stricken by poverty.

The Poorest of the Poor

The dramatic density difference only hints at disparities in living and land conditions. By the 1994 elections, Kagiso was a mess. Mines had shuttered, some leaving derelict sites behind, others attempting a modicum of cleanup. A 240-hectare slimes dam (that's nearly 600 acres) abutted the township. Sand from upwind tailings dumps presented a perpetual hazard during the windy season. When mines and other employers folded, hostels became targets for demolition. Once again dismissed as residual, residents were expelled. That's how Ramoruti and his neighbors ended up at Tudor Shaft. The settlement continued to grow over the next two decades. Neighborhoods emerged with distinct characters and names: Bull Brand, Malala, Baghdad, S-Café.

In a 2012 affidavit, Ramoruti's neighbor Phumla Patience Mjadu, an unemployed mother of four, explained that settlement residents were "all desperately poor people" with no other options. Most made less than 1,000 rands a month; those with formal employment worked as security guards, cleaners, or petrol pump attendants. Sanitation was limited to "three communal water standpipes and seven chemical toilets, which are serviced twice a week." Aside from a few streetlights, there was no electricity. Residents relied on paraffin stoves and open flames for cooking and lighting, which led to regular shack fires, especially in winter.[9] The state of their environment sickened them and made eking out a living even more difficult. Their top priority was to obtain better living conditions, starting with proper housing, complete with basic services, clean air, and unpolluted water.

The ANC's promises of a dignified life for all citizens remained a distant dream in the early 2000s. In 2009, the government launched a Community Work Program (CWP) to address the needs of the "poorest of the poor" by providing regular (if minimal) income to people in

4.2 Patience Mjadu, still from *Mail & Guardian* online video, 2013.

areas with few formal employment opportunities. Administered at the ward level, CWPs offered work two days a week, fifty weeks a year, in exchange for a daily wage. A 2016 assessment of Kagiso's program paints a portrait of township life in the words of its residents.[10]

Residents named adequate municipal services, housing, and income as their biggest needs. The absence of these basics formed the root cause of their community's most serious problems. Desperation drove young people to drugs, the most dangerous of which was *nyaope*: a hard-hitting heroin-based cocktail that could include detergent, rat poison, or methamphetamines, often with ground-up antiretroviral meds thrown in for good measure. *Nyaope* was cheap, but not free; one resident remarked that users would "take any piece of metal for recycling just so they [could] get a fix." Some lamented the levels of violence. "There are people who are raping kids in Kagiso, no one is talking about it, but it is happening," said one respondent sadly. "From last year we had an increased rate of husbands killing wives with guns," reported another. Others insisted that "Kagiso was less violent than other townships." Still others (perhaps party officials?) denied the presence of any serious violence.[11]

City leaders responded to complaints about poverty, state neglect, and pollution by putting residents (who'd been treated as waste) to work cleaning up municipal waste. The CWP did engage in activities besides waste management, including the cultivation of food gardens to nourish the municipality's schoolchildren. But most residents associated the program with trash collection, of which the most lucrative was recycling. Participants collected material that could be sold to scrap dealers; at the end of the year, the returns were distributed among the collectors. Other forms of waste work included "grass cutting, street sweeping, [and] eradicating of illegal dumps." To be clear, the dumps in question were garbage dumps, not tailings. Mine dumps lay beyond the town's administrative remit. The ANC leaders wanted quick results, not least because their Democratic Alliance (DA) opponents were gaining ground in the township. They hoped the CWP would earn citizen gratitude for the income it provided: around 540 rands for ten days of work a month, a desperately needed pittance.[12]

In short, Kagiso's ANC officials treated the Community Work Program as residual governance 2.0: the postapartheid edition.

The CWP's focus on waste work shaped how residents responded to it. Many women participants took pride in their accomplishments: "The environment is clean because of our hard work." Young men, however, refused. For one thing, they didn't want to work alongside bossy old women. For another, they saw the program as another form of waiting, another prolongation of their own "waithood," another means of delaying their full entry into adulthood.[13] They demonstrated their disdain by shouting "Clena wena Popaye!" at participants. The expression exemplified South African linguistic creativity: *clena* = clean; *wena* = you; *popaye* = Popeye, puppet, cartoon, fool. It essentially accused participants of caving to residual governance. "If you work in the CWP, it means you have lost hope in life," explained one young man. What was the point of graduating from high school if the only available work paid 540 rands a month?[14]

Women, young and old, dominated the CWP workforce. For many, the program offered a lifeline. One respondent explained, "Women are torchbearers in our communities. So, guys, males, I may not take them seriously. All what they do is to make babies and they leave babies with these people and they go. And women . . . have to make sure that their children have uniforms, that their children go to school having eaten something." Some women used the income to invest in *stokvel*, small-scale mutual aid schemes: "We all pay each other R300 which means that each member gets R3,300 when it is her turn." *Stokvel* enabled women to buy household goods and appliances; one woman reported buying a fridge. Still others invested in burial societies. Many women observed that pride stopped men from participating because they didn't want to be seen picking up "used Pampers in the street." Dealing with baby shit was women's work. Their endless, thankless, low-paid labor made the difference between full and empty stomachs.[15]

The mine dumps, however, produced pollution on an entirely different scale. Engaging that scale required resources. Kagiso residents had complained about breathing silica-laden dust since 1972. Hoping that the new constitution's declaration of the right to a clean environment would make a palpable difference, two residents had even initiated court proceedings at the turn of the millennium. But these rapidly stalled. Frustrated, people turned to NGOs for assistance. Some found support from Mariette Liefferink and her Federation for a Sustainable Environment (FSE). Others turned to the Socio-Economic Rights Institute (SERI), the Legal Resources Center, and other NGOs. In various permutations, these organizations joined forces in the fight for socio-economic and environmental justice.[16]

Allies in Action

Communities needed allies committed to refusing residuality. They needed scientists and doctors who would eschew simplistic molecular approaches and instead engage with real-world entanglements of soil, buildings, water, and bodies. Communities needed legal experts who could navigate the state and its judiciary to advocate for their rights. They needed activists who would sound alarms and stay abreast of new developments, artists who could convey the pain of residuality. They needed journalists who would keep their stories in the public eye.

It was difficult to find experts who could overcome the compartmentalizations of ordinary scientific work. Some scientists on the Rand modeled the uptake of uranium and other heavy metals in acidic water; others studied contamination in farm animals; still others measured radioactivity in soils. Such studies fit comfortably in boxes long established by scientific disciplines and molecular bureaucracies: geologists in one corner, radiation specialists in another, public health experts in a third, and so on. But compartmentalization made it difficult to understand the synergistic or stochastic effects experienced daily by affected people. And that, in turn, facilitated inaction.

Indeed, researchers who did successfully overcome disciplinary divides had difficulty gaining traction—especially if their results were disturbing. In 1996, for example, the Water Research Commission contracted geologist Dennis Toens to investigate an unusual situation around Pofadder, in the Northern Cape Province.[17] The strata there contained relatively high concentrations of uranium and arsenic—not enough to mine profitably, but enough to permeate water boreholes.[18] The head of Community Health at the University of Stellenbosch had noticed abnormally high rates of leukemia among Pofadder farmers and their employees. After collecting blood samples from over six hundred residents, doctors asked Toens to analyze mineral levels in the water. He found uranium levels to be three to four orders of magnitude higher than internationally recommended limits; arsenic levels were high as well. Combined exposure to uranium and arsenic strongly correlated with leukemia rates. Toens sent the results to the minister of water affairs with a request for "urgent intervention." He received a dismally standard response: no action could be taken, because the study hadn't considered confounding factors that might also contribute to leukemia rates. Another seventeen years would pass before Frank Winde would find Toens's article, whose title had been "changed in such a way that crucial keywords such as 'uranium' and 'leukaemia' disappeared . . . rendering any library search for these key topics unsuccessful."[19]

You couldn't evaluate the robustness or implications of a scientific study unless you could (a) get hold of it and (b) understand it. Which created yet another vicious circle. Apartheid's reliance on residual governance undermined education for generations of Black South Africans. It had instilled strong secrecy instincts within state institutions—instincts not easily shed under the new dispensation. All this made the much-touted "stakeholder engagement" even more difficult.

By the early 2000s, a host of potential community partners had emerged. The Socio-Economic Rights Institute offered legal help for the unhoused. Earthlife Africa and the Centre for Environmental Rights kept tabs on pollution and dissected plans and policies to uncover hidden consequences. The Bench Marks Foundation monitored the behavior of multinational companies (especially, but not only, mining). WoMIN focused on the impact of the mining industry on women. And more: a full list is impossible. On the West Rand, the FSE, led by the indomitable Mariette Liefferink (whom you first met in chapter 2), played a particularly pivotal role.

Liefferink threw body and soul into combating the residuality of governance. She went to just about every meeting concerning pollution in the West Rand throughout the 2010s. Subdistricts, catchment areas, municipalities, provincial authorities, and the aspirationally named

Remediation Steering Committee: each held regular meetings, with rotating casts of characters. Liefferink attended assiduously. When circumstances prevented her from showing up in person, she sent her daughter Simone, an environmental scientist. If anyone dared chide her for missing a meeting or deadline (which happened only rarely), Liefferink icily reminded them that she participated on her own time, and at her own expense. Acutely aware that she lacked scientific training, Liefferink doubled down on reading and sought out independent assessments. Her command was impressive; she didn't always get everything right, but her dogged determination to learn put others to shame. She wielded politeness as a weapon, typically thanking her interlocutors before going in for the kill, armed with a pile of evidence, a sweet smile, and devastating understatements about the "disheartening" lack of progress.

Capitalism seeks to neutralize challengers through co-optation. The beneficiaries and perpetrators of residual governance on the West Rand hoped to quiet Liefferink by bringing her into their fold. In 2009, the National Nuclear Regulator (NNR) invited her to sit on its board as a representative of civil society. Liefferink used her seat to pressure the NNR to conduct a systematic assessment of radiological contamination in the Upper Wonderfonteinspruit catchment. Halfheartedly, the regulator ran an "environmental surveillance exercise" in July 2010.[20] The team spent all of two days collecting water, soil, and food samples from ten sites that had been identified as dangerous by environmental groups. They also sent two Tudor Shaft residents for a whole-body radiation count. They gave no indication of how they'd selected those two individuals. Nor could they explain how just two cases could constitute conclusive evidence.

The analysis in the NNR's August 2010 draft ranged from sloppy, to misleading, to just plain wrong. Most egregiously, the draft claimed that because "children turn material over in their bodies more quickly than adults do," their doses would cause less harm. This flew in the face of well-established findings that children experience greater damage from radiation exposure than adults do.

Employees of the NNR calculated the potential dose to settlement dwellers as 3.9 millisieverts/year (mSv/yr), nearly four times the international recommendation for public exposure. But because the two whole-body counts showed "no internal signs of radioactivity," they dismissed the excess dose, self-righteously proclaiming that exposure didn't mean uptake. They dismissed high radiation levels at a nearby dam by smugly noting that eating the dam fish was expressly forbidden—a statement that took account neither of the extreme food insecurity experienced by many of the area's Black residents, nor of the determination of white anglers to eat their catch. Finally, the NNR's method of determining radiological risk appeared approximate at best, a fact obscured by the substantial mathematical literacy required to understand its explanation: "A risk of mortality of 10^{-6} person^{-1} year^{-1} could be deemed acceptable. This risk was further reduced by a factor of 10 to account for uncertainties in converting a release of radioactivity to a mortality risk, and *the apparent phenomenon that society becomes less tolerant of risk the more affluent it becomes*."[21] The snark about affluent society's risk tolerance seemed especially tone-deaf for the West Rand.

Activists gave the draft report a thorough drubbing, soliciting dissident voices from abroad to assess the report. These included advisors for international NGOs such as Greenpeace; Chris Busby, a British chemist who'd made a career out of challenging mainstream science; and Frank Winde, the German geographer you met in chapter 3, who'd been studying uranium contamination in the West Rand since 2002. A much-publicized story in the *Saturday Star*, titled "Living in SA's Own Chernobyl," captured international attention—especially the bit about residents having no choice but to plant their vegetables in radioactive soil.[22]

Shack dwellers learned about soil contamination through their labor.[23] David Ncwana, for example, trundled fresh soil in to fertilize his crops whenever he could find or afford it. The contrast was palpable. "Look at the difference," he told reporter Sheree Bega. "There, I mixed the soil, and you can see the mealies [maize] are green and fat. . . . Where I didn't mix the soil, the mealies are a yellow colour and small." Yet what could he do but persevere? Another resident who had been living in the settlement for seven years reported, "We're breathing in the dust. That dust is all over the furniture. In winter, when there's dust, the kids cough a lot and it's not [a] normal sounding cough. . . . When it rains [the slimes] all run in here. I'm not okay. . . . My body is always paining, and I go to the doctors, but they see nothing. Maybe it is caused by the dust, I don't know." She and her husband hoped to save enough money to return home to the Eastern Cape.[24]

Avoid Playing Outside When It's Windy

The media spotlight on radioactive tailings turned the plight of the Tudor Shaft communities into a national story and placed the NNR in the hot seat. In January 2011, the regulator asked Mogale City for an urgent meeting. As municipal officers scrambled to get up to speed, they suddenly remembered that four years earlier, their own environmental manager, Stephan du Toit, had urgently warned that these communities were at risk for heightened contamination (see chapter 2). But no one had followed up, and the problem had gotten worse, not better. The Bull Brand zone of Tudor Shaft had grown fivefold in five years, with dwellings on hollow land especially vulnerable to flooding. The Council for Geosciences and the Department of Mineral Resources (DMR) had dropped the ball. No one seemed to know who should be held liable. Municipal councillors concluded that the NNR's cursory draft report was "fatally flawed and unacceptable as a scientific document."[25]

Perhaps as a last-ditch effort to recover credibility and shed responsibility, the NNR wrote the Mogale City municipal manager, singing a thoroughly disingenuous tune: "It has become evident that the dwellers in the settlement will be exposed to elevated levels of internal and external radiation due to radioactivity in the tailings material. The NNR therefore strongly recommends that the dwellers in the informal settlement be relocated to land that is more suitable for human habitation."[26] The NNR pushed responsibility for evacuation onto the municipality. The Department of Environmental Affairs (DEA) also wanted Mogale City to cover the expense of remediation and relocation. City managers deemed such buck-passing "totally unacceptable." Mogale City would identify land on which people could resettle, but other government departments also had to pull their weight. Above all, mining houses needed to comply with regulatory requirements and pay for rehabilitation.[27]

The mayor's committee found the conflicting evidence and interpretations particularly confusing. How could it decide who was right? "Toxic environments create toxic bodies and toxic minds," it wrote angrily. Everyone had to eat and drink, but how could communities with no resources "survive and live in appalling conditions"? Especially given limited infrastructures and support systems. It was particularly shocking that epidemiological studies had been conducted on farm animals and wildlife, but not people. The "overwhelming evidence" of danger made this lack "inexcusable." An epidemiological study of cancer and genetic disease among the most exposed had become "a matter of urgency."[28]

Mogale City couldn't pay for everything. But officials did want to be seen as doing something. In February 2011, city officials contracted the Red Ant Security Relocation and Eviction Services to remove some thirty-five families whose shacks sat directly on the tailings dams, in the settlement's Malala zone. A private security company founded in 1998, the Red Ants did much of their business with municipalities and provinces. Evictions were supposed to proceed in a civil, lawful manner, and some Ants did try to treat people they moved with dignity. All too often, however, relocations were violent and chaotic, disturbingly reminiscent of apartheid-era removals.[29] Red coveralls and helmets remade day laborers (many of whom themselves dwelled in shacks) into intimidating insectoid enforcers. Their arrival never heralded good news— especially not when accompanied by a "stabilization unit" outfitted in militarized regalia.

The thirty-five households slated for removal had not received any advance notice. Making matters worse, families were sent not to RDP houses but to an area—itself subject to significant dust pollution—that had some two hundred "stands" (one-room shacks), where Red Ants unceremoniously dumped their belongings. The stands constituted a significant step down for families who'd previously lived in several rooms, and who now had to find a way to rebuild their lives with damaged goods, broken materials, and even less space.[30]

By this point, the situation in Mogale City had garnered national attention. Parliamentarians placed NNR head Boyce Mkhize on the hot seat. Officials from NNR seized the occasion to note that the methods of one self-proclaimed expert, Chris Busby, were contested by European experts and regulatory bodies. True enough: a known gadfly, Busby had achieved notoriety for his unsupported theories about radiation. Such was the risk of allyship: the scarcity of scientists willing and able to discern industry obfuscation left plenty of room for pseudo-experts to thrive. But Busby wasn't always wrong. The NNR's August 2010 draft was indeed riddled with errors. Nevertheless, Mkhize insisted that the NNR had informed locals of radiological hazards via simple, multilingual pamphlets. His testimony temporarily satisfied the committee, but the chair urged the agency to "make haste," reminding Mkhize that "the Committee was watching."[31]

The prospect of parliamentary oversight prompted the NNR to revise its assessment. Rerunning dose calculations raised the highest dose received by area residents to 4.95 mSv/yr. Other revisions paid lip service to municipal and national concerns about the health of shack dwellers but expressed no contrition. Without admitting error, the NNR now acknowledged that children were in fact "more vulnerable to radiation hazards." The revised report, issued in February 2011, made eight recommendations:

1. As far as possible, children must avoid swallowing sand when playing outside.
2. Make sure children always wash their hands before eating.
3. Do not swim in open dams and ponds.
4. Avoid using water found in dams for swimming or washing clothes.
5. Try to avoid playing outside when it is windy.
6. Avoid sweeping outside when children are nearby.
7. During raining periods make sure that mud is not carried into the dwelling area.
8. Make sure that dwellings are well ventilated.[32]

Adults the world over who struggle to stop small children from putting their hands in their mouths might marvel at the treatment of toddlers as neoliberal subjects responsible for managing their own risks. And could any South African, even the most privileged, genuinely believe that shack dwellers had easy access to running water for hand washing and laundry?

The eighth recommendation—to ensure ventilation in dwellings—should have seemed particularly odd in a report that purported to assess dust exposure. It only made sense in light of a glaring omission: as the NNR itself acknowledged, "radon measurements were not performed." Measuring radon required sealing off a space and leaving an alpha particle counter running for at least twenty-four hours. NNR scientists deemed this procedure unfeasible for informal housing. Far easier to recommend ventilation, which would enable the gas to es-

4.3–4.9 (following) The human and environmental violence of contemporary evictions often matches that of apartheid-era forced removals. When uniformed and kitted out with paramilitary gear, the Red Ant brigades become insectoid agents of terror. Yet some of James Oatway's images also capture the humanity of those recruited for these tasks; they too live in precarious circumstances, forced into accepting violence as labor in order to feed their own families. James Oatway, 2017.

4.10 Young girl playing outside her home at the Tudor Tailings Storage Facility. The FSE used this image in its 2018 annual report to highlight the absurdity of NNR recommendations. Natasha Griffiths.

cape. If any of them paused to consider the circular reasoning here—ventilation would also let contaminated dust in—they did not record any hesitation.

Evidently none of the report's readers noticed either, even though du Toit's 2007 report to the Mogale City council had drawn attention to this very issue. Du Toit had even called for an investigation into the use of radioactive mine tailings as building materials, yet another point that no one had followed up.[33] In any case, South Africa had a terrible track record when it came to admitting—let alone mitigating—radon exposure in mines.[34] No one squawked when the NNR blithely admitted that it hadn't taken radon readings. The production of ignorance feeds on itself.

Existing Exposure Scenario

Meanwhile, the NNR declared that Tudor Shaft represented an "existing exposure scenario." As you might recall from chapter 2, this meant that the International Commission on Radiological Protection (ICRP) recommendation for public radiation exposure (1 mSv/yr) need not apply. The complexity of potential exposure pathways in these scenarios, the ICRP insisted, meant that individual behavior would determine contamination levels. So anything up to 20 mSv/yr—the recommended limit for nuclear industry workers who knew the risks they incurred—

4.11-4.12 The media regularly featured photos depicting children playing on or near the mine dumps. *Top:* Young boys get ready to sled down a gold mine dump in the Jerusalem informal settlement on the East Rand, October 2014 (AP Photo/ Themba Hadebe). *Bottom:* Children play soccer in the new housing development was okay. Ultimately, said the ICRP, each nation had to determine its thresholds based on its "prevailing economic, societal, and cultural circumstances."[35]

In classic neoliberal fashion, therefore, enforceable regulatory limits depended on context, but responsibility for exposure rested with individual behavior. The invocation of "prevailing circumstances" echoed a pillar of the ICRP's long-standing recommendation that exposures be "as low as reasonably achievable," a guideline acronymized as ALARA. "Reasonably achievable" referred above all to cost, of course. This involved placing a financial value on human life. Although the ICRP denied this, it also implied that human lives had different monetary values depending on place or circumstance.[36] The 2007 iteration of ALARA placed a substantial burden on individuals to minimize their own exposures. Did residents of informal settlements have sufficient income to make meaningful dietary choices? Neither the ICRP, nor the NNR in its wake, bothered to ask.

Back on the West Rand, long-standing questions lingered. What form of remediation was "reasonably achievable"? Who would pay? Hoping for a quick fix, the NNR struck a deal with Mintails. The dump reclamation company would act as a "good corporate citizen" and spend 160,000 rands to remove the dump altogether. It would recover the costs (and then some) by remining the material.[37] Eager to wash their hands of the situation, Mogale City officials agreed. But they didn't bother informing the residents. So when the Mintails trucks arrived in late June 2012, they encountered outraged resistance from residents who knew all too well how much dust would be generated by dump removal. "The people living in the danger zone should be removed first, not the mine dump," insisted community leader Caesar Mokhutshoane. Some residents did just want the dump to disappear. One young woman who'd grown up in Tudor Shaft reported, "Our children play on that dump and it is making them sick. It's dangerous and it must go."[38] Others hoped relocating people rather than tailings would give them access to an RDP house.

Liefferink didn't trust Mintails' assurances that they would guard against dust fallout during the removal process. Joined by SERI, FSE went to court, requesting an interdiction order until compliance had been demonstrated. Nearly three hundred residents were named in the filing; state respondents included the NNR, the ministers of energy and of environmental affairs, and Mogale City. Each party insisted that responsibility lay with the others—easy to do when the tailings dam figured as "derelict and ownerless." Meanwhile, Humby notes, the NNR and Mintails portrayed themselves as "good Samaritans" willing to remove the dump despite not having a legal obligation to do so.[39]

Affidavits flew back and forth. Speaking for Tudor Shaft residents, Phumla Patience Mjadu affirmed that none were consulted about the dump removal. Quite the contrary: after the initial court filing, city officials had threatened sixty-eight households with eviction. The municipality insisted that they'd consulted the community. Mjadu flatly denied this. In mid-June, ANC ward councillor Susan Selaole had barged into the settlement with her crew and slapped yellow Xs on sixty-eight shacks without talking to any of the residents, most of whom were out scraping for food and income. The selection of shacks appeared totally arbitrary: "Sometimes the distance between a marked shack and an unmarked shack is as little as one metre."[40] A few days later, Selaole announced she'd be returning with the infamous Red Ants to conduct the relocation. Refusing intimidation, residents demanded a halt to operations until they'd been adequately consulted and a credible environmental impact assessment conducted.

The NNR retorted that marked shacks had been selected following a June 2012 radiation scan that it had performed with city officials. The Xs simply marked the most contaminated shacks. While the scan generally showed that shacks closer to the dump had higher radiation levels, exceptions occurred where residents had used material from the dump to fill spaces between their metal sheet walls. These people had only themselves to blame for their contamination. By halting the removal, furthermore, Liefferink and her allies bore the responsibility for ongoing radiation exposure.[41]

No Retreat, No Surrender

The NNR's frustration with Liefferink might seem reasonable. Officials insisted they'd reached a satisfactory solution along the lines she'd demanded. They couldn't understand why she blocked their actions. But in many respects, Liefferink's mastery of the problem exceeded their own. She knew that the act of moving dumps spread contamination, which could further sicken the very people who needed protection. She also knew the NNR hadn't assessed the radiological dangers of moving millions of tons of uraniferous tailings.

Municipal officials, meanwhile, had underestimated settlement residents. Their fury just kept growing. "We have *toyi-toyi*'d and have at-

4.13 A faint yellow X on the wall of a shack built in part from an old sign welcoming people to Kagiso Extensions 1 and 2. Mr. Ramoruti asked me to take this photo to prove that nothing had changed in the four years since the shacks had been marked. G. Hecht.

4.14 Clara Ntsepo, still from *Forgotten People* film, 2012.

tended many meetings. We get promises for new stands, but we are still here."[42] Residents expressed deep anxiety about the health consequences of their toxic surroundings, attributing their more unusual ailments to radiation exposure. They did not separate environment from housing, food, and water in making their demands, because it was the entanglement itself that attacked their bodies, lives, and livelihoods.

Some, like Jeffrey Ramoruti, had been waiting for two decades. He himself had been allocated a house in Sinqobile, a new township named after his own father.[43] But political shenanigans had resulted in his house (and others) going to other people.[44] Selling RDP housing was illegal, yet some Sinqobile units had been sold for profit. Tudor Shaft residents blamed corrupt ANC councillors who allocated the houses to their own favorites. Ramoruti's neighbor Clara Ntsepo reported that she'd complained to several authorities, to no avail.[45] Mind you, living in Sinqobile had its own hazards. The township sat next to Princess Pit, an open cast gold mine launched by Mintails in 2013 whose blasting was so powerful that it cracked house walls. Government and mine officials ignored township complaints. One resident joked that coughing "is how we say hello."[46] By early 2014, Sinqobile residents had had enough. Protests turned violent after the arrival of the police, complete with burning tires and rubber bullets. The minister of mineral resources suspended operations on the grounds that Mintails hadn't taken measures to prevent unauthorized access to the site. But two years later, Mintails obtained permission to resume by expressing repentance and promising improvement.

Back in Tudor Shaft, long-standing ANC loyalists grew fed up with their officials. Some turned to the DA, a relatively new political party and the ANC's most significant political rival at the time. Perhaps because it needed to strengthen its Black base, the DA appeared more receptive to their complaints. Tensions heightened. When Jack Bloom, a white aspirant to the premiership of Gauteng, showed up in the settlement with Liefferink and a film crew, ANC councillor Selaole tried to stop them. The DA posted a video of the encounter. Bloom virtuously signaled that he'd stayed overnight in a two-room shack inhabited by the DA's local branch chairperson: part of his "don't forget the forgotten" campaign.[47] Clara Ntsepo reported receiving a call from an ANC official, who yelled, "Yah, those white people you are working with, you don't care about us anymore. Where's the house I gave you?" Outraged, Ntsepo shot back, "What house? You gave it to somebody else! I already signed for it, but [I got] no papers."[48] Things blew up in August, when a truck roaring through the settlement killed one child and injured another. The community exploded in violent protest, which the police quelled with tear gas, stun grenades, water cannons (in a water-scarce region), and even a helicopter. One uncowed resident vowed, "I will fight until I die. No retreat, no surrender." Protestors burned Selaole's house that night.[49] Bloom did not win the provincial premiership, but the DA did increase its representation on the Mogale City council.

In a bid to pacify the parties, the DEA offered to mediate. The FSE agreed to participate; SERI did not. Nevertheless, the DEA commissioned Dawid de Villiers, a radiation protection consultant, to conduct a radiological survey around Tudor Shaft and assess potential remediation scenarios. He found significantly higher radiation levels than had the NNR, up to 6.39 mSv/yr from gamma radiation. He also managed to run radon gas monitors for forty-one days in and around a selection of twenty dwellings. These showed substantial radon concentrations: not quite at the ICRP's action level, but close enough to be worrying. Generally speaking, de Villiers displayed a more sophisticated and empathetic understanding of context than the NNR. For example, he found that external doses farther from the tailings were as high as those in Tudor Shaft. The NNR had insisted that these were simply the region's normal background levels. De Villiers countered that determining pre-mining background levels was impossible due to "years of dust deposition and pollution from mining activities."[50] Radiation levels may have been elevated for decades, but that most definitely didn't make them "normal background."[51]

De Villiers considered four remediation scenarios: barricading access to the tailings (but this wouldn't stop them sloughing off in heavy rains); capping the tailings with cement (but the bottom wasn't lined, so leaching would continue to contaminate soil and water); relocating the residents (but the housing shortage would just mean others would move in); or removing the tailings altogether. He deemed tailings removal the only reasonable solution, provided it was accompanied by frequent wetting of the area. Analysis of three possible removal tempos (10, 50, or 1,500 tons per day) showed that the quickest, which could be accomplished in eighteen days plus another five to cap off the denuded area, would produce the least overall contamination. Mintails had estimated the cost at some 300,000 rands, to which de Villiers added another 180,000 rands for monitoring and supervision. Responsibility for implementation could be shared by Mogale City, the DMR, and the NNR, with oversight by the DEA.

Meanwhile, Liefferink unearthed a 2002 legal compliance audit clarifying ownership of the Tudor tailings. Surface rights belonged to Durban Roodeport Deep, which had transferred them to Mintails. The audit had further "recommended that Mogale City Local Municipality should ensure that all surface rights owners are held accountable for any environmental issues that might arise from their action."[52] Liefferink plunked down the evidence at the June 2015 meeting of the Wonderfonteinspruit forum. A few months later, Mintails applied for business rescue—South African parlance for bankruptcy.[53]

Over the next few years, Ramoruti became a well-known face of the settlement's struggles. He appeared regularly in news stories and joined Patience Mjadu as the second named applicant in the lawsuit. A SERI attorney introduced us in 2016. Welcoming the attention of international researchers who could help publicize his story, Ramoruti took me and Tara Weinberg, who was then my research assistant, on a tour of the settlement. He pointed out shacks still marked with yellow Xs, including one that had been built using an old "Welcome to Kagiso" sign. "We are feeling taken for granted," he remarked wryly, watching us carefully to assess our reactions. The battle for housing had become fierce. Ramoruti emphasized the distinction between his Tudor Shaft community and other zones in the settlement such as Soul City, whose residents he viewed as greedy interlopers poised to snatch away his RDP allocation—assuming it ever materialized. Meanwhile, government was "eating [the housing] money."[54]

Ramoruti also introduced us to other residents. Joyce had been living in the area for seventeen years. Her current home was in S-Café, directly on the dump along with some fifteen other people. Her daughter had epilepsy, frequently triggered by mine blasting. Another resident, a young man named David Pheto, affirmed that he and others were "standing up as youth to get things moving." He indicated a small plot where a grandfather was attempting to grow mealie maize; the maize was poisonous, said Pheto, because the soil was poisonous.

Pheto and Ramoruti made sure we noticed the cracks in the ground, trenches where weeds had eroded the concrete, revealing the ruins of the former uranium mine offices below. Look down in those trenches, said Ramoruti: "Children fall in." He stopped to look me straight in the eye, making sure I had fully grasped his words. Pheto explained that he warned incoming residents not to build their shacks on these old structures. They insisted we photograph the mess before steering us onward; they wanted us to document them and their surroundings for the world to see. As we walked around the settlement, I turned my Geiger counter on. Gamma readings confirmed levels eight to ten times higher than the limit for public exposure. Our interlocutors nodded in solemn agreement. Some hung back to ask Tara about the counter.[55]

As evening fell, Mr. Ramoruti welcomed us into his multiroom shack. He showed us his papers by the light of a paraffin lamp. These included a six-page handwritten memo detailing his story and asserting, for the umpteenth time, his right to a proper house. He described an occasion in which he went with a few neighbors to city offices to (re)register Tudor Shaft residents for their RDP allocation. They were stopped by a band of Soul City residents, who forced them to abandon the quest. "If we don't comply [they're] going to burn our shacks and kill us. We complied." A few weeks later, the Tudor Shaft group tried again, "only to find another corruption in progress." That time, the city official claimed that he already had a list of all Tudor Shaft residents who'd registered for RDP houses. They grabbed his list: he'd lied. The official refused to accept their list as a replacement, and insisted they fill out a separate form for each registrant. They began doing so. But the forms ran out, leaving 348 residents unregistered. Livid, the Tudor Shaft group warned that, if necessary, they would occupy their rightful RDP houses by force.[56] They had endured more than enough waiting.

Community members and their supporters refused to be ignored. Armed with the latest scientific studies, they seized every opportunity to present their case. They kept media allies updated on their campaigns, ensuring that they remained in the public eye. Some, like Lucas Misapitso, even managed to meet individually with government officials. A twenty-four-year-old mechanical engineering student who lived in Tudor Shaft, Misapitso met with Gauteng's health minister "to tell her how these mining companies came here, made their money, and left—but don't care about the suffering." Misapitso repeated Ramoruti's refrain: Tudor Shaft was "as old as democracy," he told Sheree Bega of the *Saturday Star*, "but its very existence is a symbol that democracy has failed the poorest of the poor."[57]

In late 2016, the government sent planning minister Jeff Radebe to Mogale City for National Imbizo Focus Week. In isiZulu, *imbizo* (plural, *izimbizo*) refers to a gathering in which traditional leaders bring community members together to discuss and resolve difficult common challenges. Hoping to rescale the practice at the national level, government officials trumpeted *izimbizo* as platforms for participatory democracy. Scholars, however, observe that officials treat *izimbizo* more as politi-

4.15–4.17 Jeffrey Ramoruti wanted to be sure that we photographed the old mine offices visible through the cracks (G. Hecht), the yellow, contaminated dirt coursing through his fingers (G. Hecht), and the records he'd kept to document his struggle (Tara Weinberg).

cal mobilization tools than true listening forums; community members might raise concerns, but the government doesn't follow through.[58] Aware of this critique, Radebe promised Mogale City residents that "this *imbizo* is not just a talk shop. . . . We do not want a situation whereby we come back here next year, and listen to the same complaints that we heard today." Radebe focused much of his speech on challenges faced by residents of Tudor Shaft and Soul City (news coverage having taught him to distinguish between the two). He promised to deal harshly with corrupt officials and others who sold or rented RDP houses, "effectively depriving disenfranchised citizens who genuinely need these services." In return, he admonished, "you must always bear in mind that community facilities such as schools, post offices, multipurpose centres, and libraries that are built in your community are there for you. . . . It should never happen that one day, when you are angry because there's no water flowing from the taps, you go and torch a library, a clinic, or municipality offices. There is no heroism in lawlessness and harming the very community that you claim to fight for."[59]

In early 2017, Ramoruti and Pheto, along with some three hundred Tudor Shaft households, finally moved to RDP homes in Kagiso Extension 13.[60] A SERI attorney oversaw the process. The new homes still lacked electricity and working toilets. Nevertheless, said one resident, "it is better than living at Tudor."[61] Another 234 houses were in the works.[62] Next SERI turned its attention to advocating for housing on behalf of immigrants and others who didn't qualify for RDP homes.

Unmuting the Environment

Residents might have moved into formal housing, but the Tudor tailings remained. A growing group of Black scientists at Wits, the University of Johannesburg, and North-West University continued investigating radioactive contamination.[63] The most recent studies move past disciplinary compartmentalization, adopting research designs that seek to correlate exposure and contamination with health outcomes. A 2019 study led by Paballo Moshupya from the Wits School of Geosciences focused on radon, the most notorious element missing in the scientific story. Outdoor radon levels typically fell around 10 becquerels per cubic meter (Bq/m^3); Moshupya and his colleagues recorded levels ranging from 32 to 1,069 Bq/m^3, clearly concentrated around the tailings. Indoor levels in Kagiso maxed out at 174 Bq/m^3, above the recom-

4.18 David Pheto, Jeffrey Ramoruti, Tankodi Gala Letsang Kodi, Veli, and Joyce with her daughter, July 2016. Ramoruti's raised fist leaves no doubt about his intention to show the world that he would continue to fight for his rights. G. Hecht.

mended limit of 100 Bq/m³. They correlated their results with recorded lung cancer deaths in Mogale City, whose rates far exceeded those of neighboring municipalities. The resulting figures told a stark story. The team called for epidemiological studies to confirm and deepen the findings.[64]

Much decontamination work remained. By 2018, FSE had nineteen people on its roster, counting salaried and temporary workers, consultants, and volunteers. Liefferink had concluded agreements with Gold Fields and Sibanye-Stillwater to help fund FSE activities. Some criticized her for selling out, but she insisted that she'd retained her independence. Whatever the case, there was no sign that she or her staff had backed off. In 2018, FSE's report listed dozens of stakeholder meetings, official requests for access to information, and press interviews, as well as collaborations with seventeen other NGOs.[65] Liefferink was particularly well known for her "toxic tours," which inspired several writers and artists to pursue their own photographic or architectural projects on mining landscapes.[66] One set of filmmakers found Liefferink herself

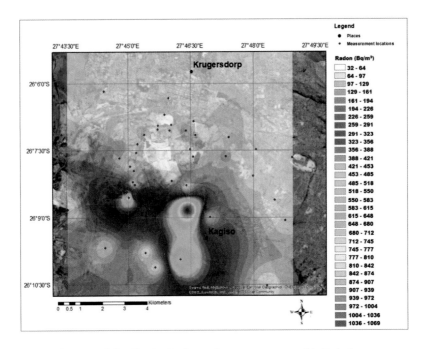

4.19 Radon spatial distribution in the study area, represented in Bq/m³. Moshupya et al., "Assessment of Radon Concentration and Impact on Human Health." Creative Commons license.

such a compelling subject that they centered her in the 2019 documentary *Jozi Gold*.[67]

Other allies also continued their pressure. In 2018, the Oxpeckers Center for Investigative Environmental Journalism launched a probe into Mintails, which had declared bankruptcy two years earlier (just as Tudor Shaft residents began moving to RDP housing). The probe deemed the Australian firm a derelict "scavenger company," one of several "under-resourced outfits that buy the scraps left over from larger mining companies and ultimately abandon them."[68] Its techniques included capitalism's time-honored shell game: multiplying corporate subsidiaries to compartmentalize profits and insulate the parent company from liability.[69] By late 2018, estimates of Mintails' environmental liability reached 485 million rands, of which only 25 million had funding.

The Oxpeckers story prompted yet another parliamentary inquiry, which found that Mintails repeatedly underreported its liability. Why hadn't the DMR required it to deposit remediation funds in escrow before the company went into business rescue? Given that the bankruptcy process had collapsed, would the DMR now hold Mintails directors per-

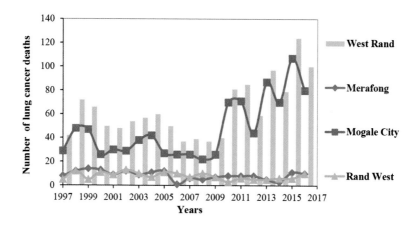

4.20 Lung cancer deaths in West Rand municipalities. Mogale City shows higher rates than Merafong and West Rand Local Municipality (encompassing Westonaria and Randfontein). Moshupya et al., "Assessment of Radon Concentration and Impact on Human Health." Creative Commons license.

sonally responsible for funding rehabilitation? Parliamentarians also disapproved of the company's shell game. Mintails traded publicly on the Australian Securities Exchange. Its South African subsidiary had had 26 percent Black ownership to comply with Black Economic Empowerment policy, but "differences" had ended that. Mintails' mining rights had been suspended due to its abysmal compliance record. Children drowned in dams that Mintails had failed to secure. Criminals disposed of bodies in Mintails-owned shafts. Despite recurring infractions, the DMR had given the company many chances to put things right. No more.[70]

Mintails abandoned its West Rand operations, leaving behind shafts, treatment plants, and other industrial debris. As soon as it fired its last remaining security officers, *zama zamas* moved in to collect the remains. Some groups were armed and demanded payment for site access. Copper cables, steel beams, office furniture, gold dust: anything with the slightest monetary value got absorbed by *zama zama* production and distribution networks. Pilfered parts included power lines to the West Rand's water neutralization plant (see chapter 2), which had to stop pumping for three months.

Liefferink denounced the danger posed by criminal syndicates to local communities, but noted that as individuals, *zama zamas* sought only "to put food on the table."[71] She aimed her strongest condemnation at government agencies. Nothing had improved since the parliamentary inquiry. In May 2019, visiting officials again witnessed "flagrant failure in duty of care, non-compliance with environmental legislation and non-enforcement of non-compliance, environmental degradation and pollution." She'd written many letters, to no avail.[72]

Out of options, Liefferink filed suit against Mintails, its numerous subsidiaries and affiliates, and a suite of government entities. Her ninety-eight-page affidavit spat fury at rehabilitation costs getting "externalised to the state, neighbouring mines, a mute environment, financially beleaguered local municipalities and communities characterised by widespread poverty, and future generations. The overall tab will be picked up by overburdened taxpayers who have little say in the ongoing corporate malevolence of the group."[73] In the very same period that contamination at Tudor Shaft was making news, Mintails "continue[d] to mine recklessly," paying no attention to rehabilitation funding. Among other actions, the suit demanded that the FSE, as a ventriloquist for the "mute environment," be included as a stakeholder in the liquidation: not for its own monetary gain, but to ensure that liquidators set aside "the financial provision for rehabilitation . . . before any of the other creditors are satisfied." A pile of photographic and documentary evidence accompanied the affidavit, along with a detailed chronology of FSE's correspondence with Mintails and relevant government entities to demonstrate the NGO's long investment in the issue.[74]

As of this writing, many dumps remain. In late 2020, Pan African Resources agreed to acquire one of Mintails' properties for 50 million rands, pending due-diligence assessments of potential profitability. Cobus Loots, the CEO, told the *Business Times* that Pan African planned to position the acquisition as a rehabilitation project. It could thus "tap into green funds in Europe" for financing, leaving its investors with "clean profit."[75] By late 2021, however, Pan African's head of investor relations was already downplaying the remediation dimension, averring only that the company would "*endeavour* to rehabilitate the surface footprints" and gesturing vaguely at "downstream benefits" for local communities. He was categorical on one point: "Pan African will not take over any material AMD liabilities as part of the transaction."[76] One can only wonder how "European green funds" will fit into this version of Pan African's strategy.

Meanwhile, the Department of Mineral Resources and Energy transferred mining rights for two Mintails properties (including Tudor Shaft) to Amatshe Mining in January 2021. December saw Liefferink once again tromping around the mine residues, this time accompanied by one of Amatshe's directors. She brought a photographer to document ongoing violations. There were many. Liefferink was distinctly unimpressed with Amatshe's rehabilitation plan, which consisted of adding lime to neutralize AMD and a vague promise to remove radioactive material. For the umpteenth time, she invoked the many reports documenting contamination; for the umpteenth time, she reminded industry that neutralizing acid mine water was only a stopgap measure.[77]

All too often, the battle against residual governance devolves into a battle of repetition and attrition. Evidence piles up, only to be "forgotten." When remediation plans gain traction, industry and state regulators introduce delays, often by invoking *feasibility*. Let's be realistic, they say. Everyone needs to take some *responsibility*, they say. Even people with no resources. Let's be *reasonable*. Let's focus on what's *achievable*.

So many words in the lexicon of residual governance refer to "ability." By which they mean the ability of industry. Or the state. Not the ability of poor people to breathe, eat, drink, sleep. Survive. Thrive.

5
LAND MINES

SOUTH AFRICANS COULD TELL hundreds of stories like Tudor Shaft's. East of Gauteng in Mpumalanga, residents live among coalfield that (still today) supply over 90 percent of the nation's electricity. Acid mine drainage (AMD) also permeates these communities. In February 2022, news broke of a colossal decant that killed thousands of fish along a 58-kilometer stretch of the Wilge River system before spilling into the downstream irrigation systems on which farmers depend. One scientist told the *Mail and Guardian* that in his entire forty-three-year career, he'd never seen anything like it. "I started crying when I saw it." Another, who'd flown over the mess to check on hippos and crocs ("they seem to be fine") explained the need to remove the dead fish before predators ate them and absorbed the contamination, propagating and biomagnifying the damage up the food chain.[1] Mpumalanga residents downwind of the province's twelve coal power plans also experience chronic contamination, forced to inhale some of the world's dirtiest air.[2] The chemical composition of contamination may vary, but daily struggles for air and water still define life for hundreds of thousands of South

Africans who strain under residual governance and struggle for what Achille Mbembe calls "the universal right to breathe."[3]

Mine lands are also omnipresent in South Africa's endless debates about land reform. Asbestos, platinum, chromium, iron . . . we could detail the toxic afterlives of South Africa's zombie mines in chapter after chapter and still not cover everything. Some six thousand "derelict and ownerless mines" festoon the country, in addition to those whose residues are owned.[4] Their leakages, debris, and emanations continue to time-bomb the future.[5] In Gauteng, planners, policy makers, and activists see the land under the piles as prime real estate, ripe for development: the key to making the city whole. This puts remediation at the center of debates about urban planning. And who better qualified than industry experts to carry out rehabilitation? As mines shut down, revolving doors spin mine officials and engineers into remediation consulting firms that profit from the harms wreaked by their own former employers (perfectly performing the second criterion for a super wicked problem: "Those who cause the problem also seek to provide a solution").[6] Ever so smoothly, industry consultants have become agents of the new apartheid.

Abandoned mine lands are land mines that detonate in slow motion. Their use as buffer zones under apartheid (and since) enacts racist separation by design, not happenstance. Attending to land justice requires reckoning with residues as spatial forms.

Spatial Injustice

In the early 1990s, urban planning activists viewed spatial injustice in South Africa as a two-dimensional affair, one that could be represented on a flat map. Mine residue areas in the city center seemed like obvious sites for democratizing urban redevelopment. Arguing that the "apartheid strip between Soweto and Johannesburg . . . holds the key to Johannesburg's door," they proposed mechanisms for transforming the territory into land for low-income housing. Activists hoped to remediate apartheid's spatial residues with urban infill strategies that housed citizens near centers of employment to reduce arduous and expensive commutes.[7]

These initial proposals did not—could not—take account of the all-important third dimension: the hollow Rand below ground and the inside-out Rand above, where the land mines in mine lands swirled.

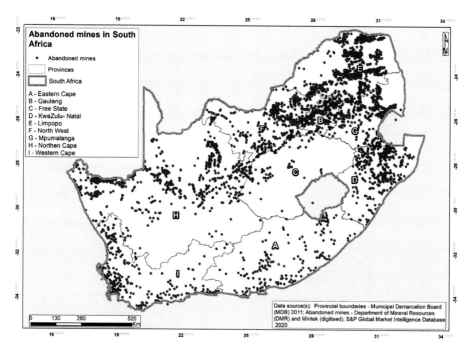

5.1 Public disclosure of mine closures by listed South African mining companies, 2020.

Mine companies and their experts held their volumetric knowledge close. Industrial secrecy—legitimated by capitalism's go-to arguments about healthy competition—was a key instrument in the politics of residual governance. Particularly useful for claiming unintended consequences, invoking confidentiality also kept under wraps geotechnical maps and expertise, making it difficult for activists to understand how underground workings limited surface development. Geographer Siân Butcher shows how this knowledge monopoly enabled Rand Mine Properties to stave off political pressure until the 1994 elections.[8]

Spatial justice remained a high priority. Black South Africans entered the democratic dispensation expecting land redistribution, restitution, or compensation. But as the years passed, very little land changed hands. Instead, historian Amanda Alexander explains, the ANC prioritized private property rights. This could be seen in the party's approach to housing provision, which was influenced by the World Bank officials who courted the ANC in the lead-up to 1994. Bank bureaucrats argued against a supply-side solution to housing using new construction, insisting it was wasteful. Instead, they propounded a demand-side model for housing and services to leverage market efficiencies. Cost recovery principles would enable infrastructural development to pay for itself. Poor people could finally conceive of themselves as consumers.[9] By treating the perpetually amorphous land question separately from urban housing policy, Alexander argues, the ANC only aggravated the dispossession of Black South Africans.

Still, the ANC government did end up engaging in substantial supply-side construction via its Reconstruction and Development Programme (RDP). The housing need was too vast, the demand too obvious. In its haste, however, the program located many developments alongside dumps and dams. People continued migrating to the region in search of livelihood, so demand endlessly outstripped supply. As the struggle to overcome residual governance gathered steam, municipal planners and developers hired consultants to assess the consequences of building on or near dumps and dams.[10]

By 2009, mine residues had seized the headlines. Provincial authorities could no longer ignore their volumetric violence. A first study commissioned by Gauteng's Department of Agricultural and Rural Development (GDARD) concluded that of the 321 square kilometers covered by Gauteng's 374 mine residue areas (MRAs—by now this had become a bureaucratic term of art), the pervasive presence of radioactivity meant that only 25 square kilometers could be rehabilitated at low cost. How could more of this land be liberated? Numerous obstacles prevented meaningful calculation. Jurisdictional confusion reigned. Severe undercapacity in state agencies, with up to 30 percent of positions unfilled, compounded the problem. A woefully inadequate monitoring system (if *system* was even the right word) didn't help.

Buffer zones around radioactive dumps remained essential. Regardless of potential monitoring improvements, GDARD warned that siting low-income housing on abandoned mine lands would "expose the poorer sector of the population to grave health risk."[11] Reclamation should focus instead on turning mine residue areas into sites for new industrial activities. An aspirational decision tree promised streamlined decision making, appearing to offer choices while remaining politically palatable to mine magnates who craved walk-away solutions.

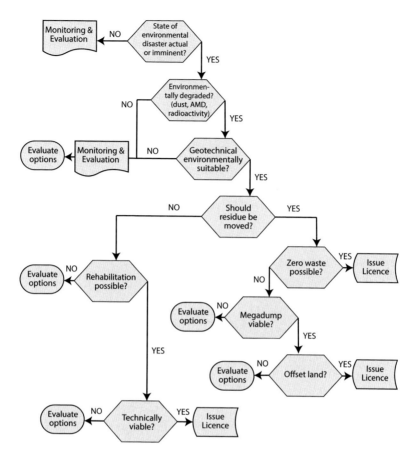

5.2 Gauteng's MRA decision tree promises a streamlined operational flow for mine rehabilitation: a generic series of yes/no decisions that apply everywhere, smoothing out pesky complexities and site specificities. GDARD, "Gauteng Mine Residue Area Strategy," 2012.

City-Region

The complexity of spatial injustice required redress and remediation on a regional scale. In 2006, the Gauteng Provincial Government began contemplating a new designation for the conurbation: the Gauteng City-Region, or GCR.[12] Coined by American urbanist Jane Jacobs, the phrase *city-region* described a space in which a city formed the nucleus of a larger area, whose coherence depended on "economic energy" rather than geographic boundaries.[13] At the turn of the twenty-first century, urban geographers focused their attention on global city-regions: places that "function as the basic motors of the global economy" while also being shaped by local forces.[14] As an organizing concept, the city-region enabled a reworking of geographic and political scales, eschewing neatly nested hierarchies in favor of interpenetration. Gauteng has certainly served as a motor of the global economy. Its governance has operated on a regional scale since the 1902 establishment of the Rand Water Board.[15] Ministries, provincial authorities, and national agencies engaged with Gauteng localities at a variety of scales, frequently pushed or pulled by community protest, activist demands, and the vagaries of party politics and interpersonal conflict. All this has made for extraordinarily complex governance configurations.

The entanglement embraced by the city-region concept invited fresh approaches to apprehending Gauteng and its governance dilemmas. In 2008, a new think tank rose to the challenge: the Gauteng City-Region Observatory (GCRO). A partnership among the Gauteng Provincial Government (which provided core funding) and the city's two largest universities (which offered in-kind support), the GCRO was expected to establish strong links with "all the higher education institutions, as well as knowledge councils, private sector think-tanks, research NGOs and information-exchange and learning-networks operating in the city-region." Staffed primarily by social scientists—some from academia, others with extensive government experience—the observatory walked a fine line. Only total transparency and intellectual independence could guarantee its legitimacy. But it also had to produce "on-request policy support" in the form of commissioned studies and data sets, short- to medium-term projects, and networking facilitation. The mandate for GCRO was to develop the "strategic intelligence" essential to making governance "more functionally integrated, spatially coherent, economically competitive, creative, innovative, environmentally sustainable and socially inclusive."[16] A tall order.

The political and epistemological sophistication of GCRO researchers was evident from the start. Elsewhere, urban geographers deployed the city-region concept as a simple descriptor, using stable criteria to establish whether a conurbation was in fact a city-region. But GCRO's savvy social scientists understood that the relation between signifier and signified was far trickier. As they saw it, "the 'Gauteng City-Region' is both an *actually existing place*, holding more than a quarter of South Africa's population, and a *political project* for better government and governance in this all-important part of the country."[17]

Done right, therefore, describing the city-region could help bring it into being as a coherent object of governance. Researchers approached their task as a collaborative experiment, one that performed some of the interpenetration that it also studied. Rather than imposing scales in advance, for example, they asked questions like "At what scales do residents of Gauteng imagine their 'community' to be? What kinds of social boundaries do they draw around their homes, neighbourhoods, social networks, cities, the province, nation and at other broader scales such as the continent?"[18] The choice of what to describe had political consequences. Omissions mattered as much as commissions, the GCRO knew full well. Failing to produce a data set or research a topic could have a deep impact on people's lives.

In essence, the GCRO's job was to repair epistemologies of ignorance. Capturing the city-region in all its complexity and contradiction had to precede any plan for imagining its future. Researchers sprang into action with both quantitative and qualitative research. Quality-of-life surveys rapidly became their signature product. These covered demographics, basic services (water provisioning and refuse removal occupied center stage), food insecurity, race relations, migration patterns, debt, schooling, access to emotional support, and more. "Maps of the month" typically highlighted some aspect of their findings, and the corresponding data sets were freely available online. In these and other ways, the GCRO supported aspirations to social justice, as well as the realization of human rights promised by the country's constitution.[19]

Gauteng's mine residue areas already occupied center stage when the GCRO swung into action. It launched its Provocations series by inviting Terrence McCarthy, a geoscientist at the University of the Witwatersrand, to write about AMD. His paper exemplified the series' commitment to publishing controversial ideas—so much so that his views on who should pay for AMD remediation seemed at odds with the GCRO's own social justice priorities. McCarthy proposed pumping water out of the hollows in perpetuity, mostly with funds from "the national fiscus" supplemented by lesser contributions from still-active mines and local and provincial governments. "Clearly," the geoscientist insisted, defunct mines "cannot be held responsible (and financially liable) for footing the bill. In reality, taxes on past mining activities have benefitted all of us—the infrastructure we enjoy in the GCR has been funded in no small part by mines that are now defunct."[20]

Things seemed off to a rough start. Who was this "we"? Apartheid's infrastructure, after all, had been designed to create and maintain inequality, to separate "us" from "them," to prevent the emergence of a meaningful "we." Those comfortably ensconced in the gated communities of the northern suburbs might laud apartheid's infrastructural legacy; informal settlement residents trudging long distances to communal water taps and toilets, not so much. The GCRO's own preface to McCarthy's provocation evinced discomfort, reminding readers that "civil society activists believe that mines have enriched themselves without any acknowledgment of the costs to the environment and are continuing to enrich themselves while doing little for the environment—damage is a 'cost of business.' Many believe that a policy of 'the polluter pays' should apply." Still, the fact remained that truly defunct mines—those without solvent corporate descendants hidden via legal shell games—couldn't be held to account. Ultimately, government had to prevent operational mines from "sidestep[ping] culpability for the destruction of the environment because some of their predecessors are not around to foot the current bill." The state and existing mines had to work together to address the situation.[21]

Clearly, the analysis needed to go deeper. In 2013 the observatory launched a multiyear project called Mining Landscapes of the Gauteng City-Region to assess the challenge to social and environmental justice posed by the Rand's hollow, inside-out topography. Geographer Kerri Bobbins began by reviewing how acid water threatened both the quantity and quality of the region's water supply. The industry had long externalized environmental costs, and the persistence of poor regulation failed to hold companies accountable.[22] One example among many: although legislation prescribed a 500-meter buffer zone around mine residue areas, no one enforced these exclusion zones. Tens of thousands of people eked out a living at the base of the dams while radioactive mine effluents polluted their water. Bobbins illustrated this confluence of issues with a striking map.

Bobbins also highlighted the lack of public consultation surrounding the design and construction of the water treatment plants. The suspension of the environmental impact assessment (EIA) process for the Central Basin plant, she argued, set a bad precedent. Government had a responsibility to empower the public by providing good information in a timely manner, but it had de facto delegated this task to private sector consultants. Nongovernmental organizations had to deploy the Promotion of Access to Information Act to obtain anything meaningful. Information, if shared, was delivered at public ward meetings. But the GCRO's quality-of-life survey showed these meetings reached fewer

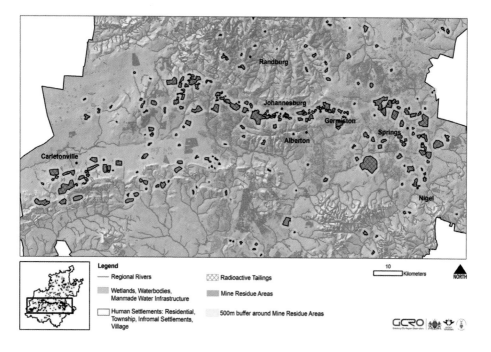

5.3 The relative locations of water bodies, radioactive tailings, mine residue areas, and human settlements in the GCR, 2015. Bobbins, "Acid Mine Drainage."

than 13 percent of municipal residents, who preferred newspapers, pamphlets, and radio as sources of information. The costs of AMD mitigation also caused concern: estimates ranged from 10 to 30 billion rands. Bobbins concluded that AMD and its mitigation threatened the democratic processes meant to create a new South Africa.

The final project report in 2018 brought mining landscapes and their inhabitants to life. One of the most gripping and innovative sections was a photo essay commissioned from artist and urbanist Potšišo Phasha, based on work he'd completed five years earlier. Titled "Scavenger Economies of the Mine Dumps," the essay followed a group of mostly Zimbabwean men who combed mine residue areas for abandoned machinery and scrap metal. Phasha's images portrayed these men from the rear, or from afar, or—if close-up—only their working limbs, because showing faces could further imperil their already precarious existence. His accompanying text explained how their work intersected with other informal networks, from the food sellers who nourished them to the criminal gangs who preyed upon them.

5.4–5.6 Potšišo Phasha, "Scavenger Economies of the Mine Dumps," reprinted in Bobbins and Trangoš, *Mining Landscapes of the Gauteng City-Region*. This project began for Phasha while a student in urban planning, when he was curious about the men working on the dumps and skeptical of their standard portrayal as criminals. The original project, presented in color for GCRO, sought to document the lived experience of these men without placing them in danger by revealing their faces.

Phasha showed how scavenging intertwined formal and informal economic practices. On one dump, the men worked in the wake of a gigantic excavator, collecting scraps unearthed by the machine. The constantly changing morphologies of the dumps regularly presented new opportunities and dangers. At the day's end, the men sold their findings to a scrap dealer. Prices ranged from 1.60 rands for a kilogram of light steel to 50 rands for a kilogram of copper. Some scrapyard owners cultivated the loyalty of scavengers by hosting *braais* (barbecues) and providing soap and water for them to wash up after a long dusty day's work. The calculator used to tally earnings, Phasha wrote, "stands as a tool of governance, an important link between the formal and informal economies. It subjects the men to another world of rules and logic, a numerical system that makes the thrown-away metal fragments they collect . . . materially significant."[23] The commissioned essay testified to the GCRO's commitment to include all the agglomeration's residents in its appraisals and recommendations, and to do so with a range of epistemological approaches.

Is This Who We Are Now?

The place of Phasha's work in the GCRO report highlighted the importance of artists in shining a light on the predicaments produced by residual governance. In the report itself, the photos served a documentary purpose; accordingly, they appeared in color. But the project—and especially its academic title—had left Phasha uneasy. Labeling the men "scavengers," he worried, "diminished the human aspect of what was unfolding on Johannesburg's mine dumps."[24] In the years since 2013, when he took the photos, *zama zamas* had become the most demonized denizens of Gauteng, easy prey for criminal gangs seeking to conscript their labor and a ready target for xenophobic violence. Humanizing them became an increasingly radical act, more so with each passing year. In 2019, Phasha reprised the collection, this time working in sepia, a "colour palette vivid in my childhood memory." Sepia invoked how he—like the men he photographed—had been "raised in the soil" and thus saw "the urban environment with those same eyes."[25] Now titled *A City on a Hill*, the new iteration of the project weaves poetry into Phasha's image tapestry in the form of formal verse and creative captions. One sequence portrays men lugging their findings to a scrapyard.

Phasha's refusal to endanger his subjects by showing their faces makes it difficult to invoke their thoughts and feelings. The captions fill the gap. *Plot Twist* invites you to consider the small but steady indignities involved in ferrying heavy materials with makeshift equipment, while reminding you that these men have purpose. They're writing their own plot during their time *Between the Womb and the Tomb*. They *Knowhere* they're going as they navigate the flow of the city, building their own *Bridge over Troubled Water*. And how troubled that water is for them! The media denounce them as thieves, especially when stripping copper, the most lucrative of the recovered metals, from electrical cables. Or they're accused of being *izinyoka*: snakes who steal components or power from the electric grid.[26]

Phasha's longer poems introduce additional protagonists. Written from the perspective of a "rich young ruler," "Load Shedding" (the South African term for rolling blackouts) begins with a synesthetic treatment of Johannesburg's inescapable contradictions, where infrastructures and residents throb with residues of many pasts:

> *Gloom stalks the city tonight.*
> *The darkness is extra bold*
> *When you're known as the city of gold.*
> *But the gold is gone, and so are the lights.*
> *Thick pollution in the air drowns the moonlight.*

Phasha explained to me that the protagonist is "struggling to come to terms with the consequences of some of his corrupt dealings as they manifest through the urban environment and broken urban governance systems." Load shedding itself takes on metaphysical dimensions, evoking how the human condition "unfolds when we walk in spiritual darkness instead of being illuminated from within." Seeking to shed his own load, the rich young ruler watches a woman make her way home and wonders:

> *Should I give up my seat on the gravy train for her as a gift?*
> *Maybe there's a chance she won't see me as her rival.*
> *Will she forgive me for my sticky hands*
> *That stole her chances of survival?*
> *She might agree my greed left her children scrambling for crumbs*
> *   up in the hills.*
> *I salute her strength and all she's doing to pay the bills.*

5.7 *Plot Twist*

5.8 *Between the Womb and the Tomb*

5.9 *Knowhere*

5.10 *Bridge over Troubled Water*

5.7–5.10 Potšišo Phasha, *A City on a Hill*, 2021. In his artistic rendition, Phasha presents his work in sepia tones, matching his childhood memories of rural Limpopo. The dumps are in central Johannesburg; Phasha's frames convey economic and environmental continuity between rural and urban, blurring boundaries often kept separate by urban planners. Mine dumps themselves blur these boundaries.

5.11 Potšišo Phasha, *Piece of Mined*, from *A City on a Hill*, 2021.

Oh, the allure of a world in which young Turks would understand the violence wrought by their "sticky hands," a world in which they'd want forgiveness!

> *Yet, I am the thief in this city and I know why.*
> *I stole her shine when I turned a blind eye.*
> *I crushed her future when I made decisions with my stomach.*
> *I squeezed the fruit of her sacrifice*
> *And milked her rich treasures to her demise.*

The words land with even more force when Phasha recites them in the short film that accompanies the poem, his voice rich with questions. The young ruler's verdict on himself applies well beyond Johannesburg:

> *Oh, what divine punishment I deserve*
> *For killing your radiance and draining your reserves.*

Phasha expands this sense of transcendence in "Forever." The poem immediately follows *Piece of Mined*, which shows men laboring to extract a metal pipe. It explores how the exigencies of the present blend with reveries of the future:

I guess this is who we are now.
We are far beyond reach,
Living in the future way ahead of our time.

The *we* here is expansive. It includes the *we* who "built mansions on the information superhighway," the *we* who "bear the cost of expansion and growth," the *we* who "seek to be everywhere and do everything all at once," the *we* who "strive to live forever," the *we* who want to "own forever," the *we* trapped by the allure of a permanently changed future, seduced by an elusive *forever* that is eternally in the future.

Today, forever starts tomorrow.
Tomorrow, she dawns the day after.

Reaching beyond southern Africa, the poem is also a lament for the planet. Which forever will tomorrow bring?

Megaprojects!

A key bit of code in apartheid's still-churning algorithm, spatial injustice was exacerbated by the immense housing backlog, estimated at 2.1 million units nationwide in 2018. Twenty percent of Gauteng residents lived in informal housing, the highest rate in the nation.[27] In the hands of politicians and policy makers, spatial injustice served as a proxy for economic, environmental, social, land, and other injustices.

Consider the ongoing debates over urban form, which pit advocates of urban infill against proponents of satellite settlements. The urban infill camp experienced a major setback when Lindiwe Sisulu, the minister of human settlements, announced in 2014 that the national government would no longer pursue small-scale housing developments. Henceforth, all subsidized housing would consist of megaprojects of at least fifteen thousand homes. "Nothing short of a total mobilisation of society around the issue of the provision of housing for the poor will solve the problems we confront in the short term," she explained. The ANC had been in power for twenty years. "We can't go into 30 years of freedom with a huge backlog."[28] Sisulu promised six million units within five years. The initiative would work in concert with the Infrastructure Development Act, recently signed into law. One unspoken motivation for action: the upcoming general elections. Tudor Shaft's

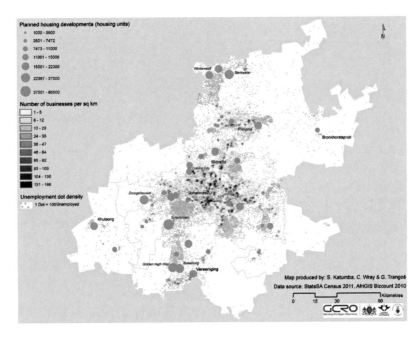

Planned housing developments (housing units)
- 1000 - 3500
- 3501 - 7472
- 7473 - 11000
- 11001 - 15000
- 15001 - 22366
- 22367 - 37000
- 37001 - 60000

Number of businesses per sq km
- 1 - 5
- 6 - 12
- 13 - 23
- 24 - 35
- 36 - 47
- 48 - 64
- 65 - 82
- 83 - 103
- 104 - 130
- 131 - 198

Unemployment dot density
- 1 Dot = 100 Unemployed

Map produced by: S. Katumba, C. Wray & G. Trangoš
Data source: StatsSA Census 2011, AfriGIS Bizcount 2010

5.12 GCRO Map of the Month, May 2015, showing the proposed locations of Gauteng's housing megaprojects. https://gcro.ac.za.

Jeffrey Ramoruti was far from alone in his frustration and rage at the ANC party operatives who had left him unhoused for over two decades; Gauteng had over 687,000 housing applicants.[29]

The ANC won the elections in May 2014 with just over 62 percent of the vote nationally. But this was down from nearly 66 percent in the 2009 elections. In Gauteng, the ANC dropped below 54 percent of the vote, losing seats to both the Democratic Alliance and the newly formed Economic Freedom Fighters. Hoping to fend off further damage, provincial authorities launched their own housing megaproject plan in April 2015. A few weeks later, however, a GCRO map showed that projects were primarily positioned on the periphery. Their prospect of transcending apartheid spatiality looked grim.

The megaproject dream reflected long-standing government enchantment with blank slates for social transformation, dating to the heyday of colonial urbanisms.[30] Geographers Richard Ballard and Margot Rubin write that Gauteng authorities hoped that new cities would "design in a greater degree of social and economic integration."[31] David Makhura, the province's premier, warned that private developers needed to include low-income housing in their plans: "We want inclusive development, not enclaves of the rich." Nevertheless, he promised developers that Gauteng's Human Settlements Department would reduce the turnaround time for EIAs to three months; government must not "destroy the passions that the private sector partners have," lest they "lose their appetite and go elsewhere."[32] Residual governance thus served the ANC government in much the same way it did its apartheid predecessor.

Expedited EIAs could perpetuate environmental injustice. The Fleurhof megaproject in Soweto—over ten thousand units located 7 kilometers west of Riverlea along the Main Reef Road—presented a case in point. The project may have (in the proud words of its developer) won "various prestigious awards," but mine dumps loomed on all sides.[33] Still, people desperate for adequate housing were eager to move in— even those acutely aware of the dangers posed by the dumps. Activist Charles van der Merme, for example, told a journalist he'd "move there in a flash." He'd been living with his in-laws in Riverlea for years. "I applied for a government house in 2000 and I'm still homeless. . . . Despite all the ills from the mine dumps, I would like a house of my own. Where else can I go?"[34]

Megaprojects made excellent political fodder for the Democratic Alliance. Writing in the *Sowetan*, Gauteng DA leader John Moodley compared the schemes to those of Hendrik Verwoerd, the architect of grand apartheid. Both Verwoerd and Makhura justified "housing the Black population in specific, purpose-built mega-townships" by appealing to the cost savings of scaling up. History had begun to repeat itself, with "former liberation movements [beginning] to mirror the behaviour and policies of the former oppressor." Moodley criticized existing RDP housing developments for their isolated siting and substandard services. By contrast, he noted, under a DA-led coalition government, the city of Johannesburg had increased the budget for purchasing inner-city buildings and refurbishing them as low-rent housing for 1,164 families. Makhura, claimed Moodley for the DA, was more interested in ribbon cutting than in community needs.[35]

Ballard and Rubin argue that the megaproject policy was undercooked. The national government didn't come close to meeting its 2019 target for six million units. Ribbon cutting could only proceed by declaring already-existing developments to be megaprojects. Lufhereng, for example, was proposed in 1997, launched in 2008, and absorbed into the megaproject portfolio in 2015. Located on the far western edge of Soweto, Lufhereng promised its residents economic opportunity in the

5.13 A boy in Soweto shields his eyes from dust flying off a nearby mine dump, 2011. Samantha Reinders.

agricultural sector. This orientation responded to the geotechnical conditions of the hollow Rand, which could not support surface-level heavy industry there.[36] Although Lufhereng itself sat outside the 500-meter buffer zone around mine residues, the agricultural land designated for its residents abutted one of those dumps. Promoters mentioned neither this nor the lack of easy access to major public transportation routes. Rather, they touted a mixed-income community, an imaginative architectural approach, and a host of services and amenities. Their plan would even obviate informal economic activity: Lufhereng, they insisted, would not be tainted by backyard shacks, *spaza* shops, or home hair salons. As any architect or historian of urban Africa could readily have predicted, these businesses emerged anyway.[37] Sarah Charlton sees them as evidence of success, rather than failure: signs of a healthy community.[38]

Clearly, megaprojects didn't necessarily offer escape from the land mines of residual governance. The mine lands were simply too extensive for any single housing strategy to overcome.

Infill

Gauteng's provincial and municipal officials had little choice but to go along with the national government's megaproject commitment. Their acquiescence didn't necessarily exclude alternatives; megaprojects and infill were not inherently incompatible. Calls persisted to "stitch the city back together." The vaguely nostalgic phrase implied that Johannesburg had once been whole. In reality, of course, it had been born divided.

Regardless, the city-region needed purposive suturing. Writers described Johannesburg as elusive, restless, anxious, frightened, panicked, addictive, extreme, haunted, bloated, agonistic. They also celebrated its vibrant diversity. This was a place where Muslims met Methodists, where Somalians and Ethiopians made community, where some Black South Africans established themselves as middle-class property owners and consumers while others picked waste outside gated communities. It was a place of defiance; of state violence and intimate violence; of fashion, art, music, and literature; of transience and permanence; a place where architectural innovation was expressed in the stunning design of the Apartheid Museum as well as in the shacks that many made their homes. Johannesburg consistently held the record for most unequal city in the world, but there was so much more to the city than extreme contrast. Infill meant taking all this into account when imagining and planning the future.[39]

Infill also meant tackling the mining belt. Urban planners and architects took up the challenge. Landscape architects, with their knowledge of organic forms, were particularly well equipped to explore spatial transformation under conditions of volumetric violence. Their proposals offered no easy outs. Like others who approached this super wicked problem honestly, they concluded that permanent mine closure—the walk-away solution craved by industry—was a pipe dream. In the real world, repair and remediation, if they occurred at all, had to be approached as an ongoing process rather than a finite goal.[40]

Plan upon plan emerged, sometimes overlapping, occasionally at cross-purposes. The most recent (as of this writing) included the *Spatial Development Framework 2040* (*SDF 2040*) developed in 2016, and the subsequent *Nodal Review Policy* in 2020. Both responded to the 2015 Spatial Planning and Land Use Management Act, which spelled out five development principles for the nation: spatial justice, spatial sustainability, efficiency, spatial resilience, and good administration.[41]

Via *SDF 2040* and related efforts, planners sought to translate these principles into strategies for densifying the inner city and other economic nodes, creating effective and affordable public transit systems, and improving the quality and quantity of services and housing for the most marginalized residents. The proposals unequivocally favored infill brownfield development over peripheral greenfield development.

Home to 40 percent of the city's population, Soweto held the key to this strategy. Planners wanted to make it a vibrant economic center solidly linked to the central business district (the transportation part) and also a place where people could live near their employment (the densification part). "This approach," they promised, "will also address the remaining areas of deprivation within Soweto."[42] In stark contrast to the national megaproject vision, which seemed determined to eliminate all informal housing, *SDF 2040* sought to integrate "backyarding": homeowners renting out shacks or concrete block rooms erected behind their houses. Building on a previous program aimed at ensuring minimum health and safety standards in these self-built structures, *SDF 2040* saw backyarding "as part of the housing solution."[43] So did residents who bought bricks from the small-scale vendors you met at the beginning of this book.

The real key to making Soweto "a true city district," however, lay in "unlocking the mining belt," which planners agreed was "the most prominent feature of urban fragmentation in the city." Rand Mine Properties, now going by the moniker iProp, retained its status as one of the city's largest landowners. It promised planners that some 1,200 hectares of mine land could potentially be rehabilitated for other uses. Roads could transect the belt. Business and residential developments could occupy remediated land. An open space system could permit gradual rehabilitation of more intractably contaminated areas for far-future development. All this needed to take the belt's pollution into account; remediation would take several decades and require broad stakeholder participation.[44] Complex and time-consuming, but doable. Neither *SDF 2040* nor the subsequent *Nodal Review Policy* offered implementation details, however. For those, you had to consult a separate document: the Strategic Area Framework (SAF) for the West Mining Belt.

By now you probably recognize the pattern. Lacking in-house expertise, the city contracted the SAF job out to a private consulting firm, Plan Associates. The firm harked back to the mid-1960s heyday of grand apartheid. Its website proudly described the initial partners (a town planner, a land surveyor, and a statistician—white men one

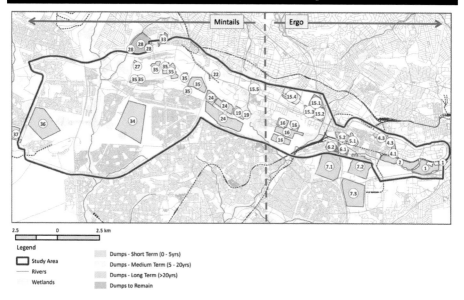

Mintails | Ergo

2.5   0   2.5 km

Legend

☐ Study Area
— Rivers
   Wetlands

   Dumps - Short Term (0 - 5yrs)
   Dumps - Medium Term (5 - 20yrs)
   Dumps - Long Term (>20yrs)
   Dumps to Remain

5.14 This map from the 2016 Strategic Area Framework showed which dumps should be handled by Mintails, and which by ERGO. The neat division conveys the impression of smooth, clean removal operations. Pretorius, "West Mining Belt."

and all) as having "recognised the need for a 'one-stop-shop' service, consisting of town planning, engineering services, land surveying and architecture." The neutral drone of the list drowned out the violence required to perform those services, instead laying down a beat for a narrative about smooth democratic transformation.[45]

Plan Associates prepared an SAF that outlined the steps needed to transform the "development vacuum" of the belt into usable territory. Impressively packed with detailed color maps and ambitious renditions of a post-transformation mining belt, the SAF brimmed with optimism about the city's long-term future. Scratching beneath the surface, however, showed that the authors lacked substantive understanding of the technical and biological dimensions of the task. They listed but did not assess AMD treatment technologies. Radiation barely came up, and their citations did not indicate mastery of the scientific or technical literature. Their faith in dump vegetation ignored the long history of obstacles, setbacks, and failures that had beset previous attempts.[46]

Unlocking the mine belt would require the city to form public-private partnerships with the remaining mining companies, said the consultants. Government (the scale of which remained largely unspecified) had to provide capital, licensing, environmental approvals, and a variety of services to smooth the way. It would need to collaborate with Mintails and ERGO in plotting the roadmap to reclamation. Rather than providing the key to unlocking the mine belt, the SAF sent city officials to the remining companies to negotiate access conditions.[47]

Such deference only buried the mine lands deeper in the code of the new apartheid. In this case, the risks of privatization immediately became apparent. A few months after the consultants rendered their report, Mintails entered liquidation. By 2018, it had posted a reclamation debt of nearly half a billion rands.

Mine lands were too toxic for infill policies to neutralize their noxiousness, just as they'd been too spread out for megaprojects to avoid altogether. The infill proposals especially highlighted the super wickedness of the residues. The very entities that had caused (or severely aggravated) the problem—from consultants who'd built their business under apartheid to companies that had profited from cavalier residue removal—were being called upon to fix it. The long history of residual governance had written this paradox into the terms of the racial contract and the technopolitics of racial capitalism.

Paying for Perpetuity

Mining long figured as the nation's lifeblood. The heyday of gold had passed, but many other minerals awaited potential excavation, most notably platinum. Could mining proceed without time-bombing the future?

The question broke down into several parts. Could mining shed its apartheid algorithm? Could South Africa develop a stronger regulatory regime than the minimalism and delay of residual governance? What would such a regime look like? How would it be enforced? South African officials began trying to reform mine legislation in 1991. Three decades later, the effort is still ongoing, all too often held hostage by industrial interests that insist on (impossibly) permanent solutions so that they don't have to pay in (or for) perpetuity.[48]

By the early 2000s, at least eight major pieces of national legislation applied to mine closure and rehabilitation, including the Minerals Act (1991), the Environmental Management Act (1998), the National Water Act (1998), and the constitution itself. Some underwent amendment in subsequent years, while others were replaced entirely. Various provincial and municipal legislation also applied to mines, as did a long list of more specific regulations, frameworks, and guidelines (over fifty, according to one count).[49]

This proliferation of legal instruments testified to how deeply mining infused the South African state and its body politic. It also generated profound confusion. Some bits of legislation overlapped with other bits. The power to adjudicate contradictions remained contested. Vitriolic jurisdictional battles ensued, often pitting the Department of Mineral Resources (DMR) against the Department of Environmental Affairs. On some occasions, state agencies were only too happy to pass the buck. At other times, they refused to relinquish control.

This confusing complexity facilitated loophole exploitation.[50] All too often, the legal designation of "care and maintenance," meant for mines not currently under operation but still tended by their owners, provided political cover for abandonment and dereliction.[51] According to one estimate, barely a handful of South Africa's 2,900 closed mines complied with the requirements for a closure certificate.[52] For example, an operator could place a mine under indefinite care and maintenance by characterizing the mine residues as potential resources rather than waste. Another popular work-around, as noted in chapter 3, was for white-owned companies to sell their liability-laden properties to "junior" Black operators. After mining the dregs for profit, smaller companies would be driven into bankruptcy by the much larger mess made by the previous owners. Both strategies were perfectly legal. Both further entrenched the new apartheid. Neither addressed the legacy of pollution. More often than not, they made it worse.

Not wanting residual liability to hamper future growth, the industry still yearned for walk-away solutions. Closure certificates were complex and costly, but they could offer a version of walk-away by releasing companies from future liability. That incentive disappeared after 2012, however, following a court case against Harmony for its attempt to deny responsibility for the mess it had sold to Pamodzi (see chapter 3). Because it no longer owned the land, Harmony argued that it no longer bore responsibility for its AMD. But Judge T. M. Makgoka of the North Gauteng High Court would have none of this. If an entity "sever[ed] ties with the land fully knowing that [their] validly imposed obligations cannot be fulfilled," Makgoka wrote, they "can hardly complain if it is insisted that [they] should comply with those before [they are] discharged of them."[53] This judgment laid the groundwork for another round of legislative amendments holding previous owners responsible for environmental liabilities regardless of closure certificates.[54] Tracy-Lynn Humby describes this outcome as "the spectre of perpetual liability."[55] New laws sought to minimize time-bombing by requiring permit applicants to plan rehabilitation financing, setting funds aside throughout their operation's lifetime so that these funds would be available after the company ceased to exist.

Harmony had clearly deployed underhanded tactics to duck responsibility. But the problem was systemic. The management and treatment of AMD on the Rand could last for decades, possibly centuries—well past the likely lifetime of any corporation. In the face of mounting frustration at regulatory intricacy, state officials began contemplating a more streamlined approach. They hit on an arrangement to split power between mineral and environmental authorities. Known as One Environmental System, the arrangement went into effect in December 2014. It specified that mining be regulated by the nation's environmental act. Environmental Affairs would set benchmark standards, but Mineral Resources would issue the permits. The minister of environmental affairs would adjudicate appeals. "Fixed and synchronized timeframes" would govern environmental and social impact assessments and their resulting authorizations.[56]

The negotiations leading to this arrangement gave considerable voice to business interests, but very little to communities. "Synchronized timeframes" really meant shortened EIAs.[57] Civil society watchdogs strenuously objected. "Some environmental impacts cannot be assessed in days and months," seethed the Centre for Environmental Rights. "Think about our great seasonal rivers, which flood in summer and run dry in winter. Or think about how levels of dust . . . in the air change in Gauteng and Mpumalanga from winter to summer." The three-hundred-day EIA process imagined by the agreement was woefully inadequate; the twenty-day appeals period, even worse. While the Centre for Environmental Rights and others had "for years argued that the environmental impacts of mining should be regulated under environmental laws just like all other industries, the political sacrifices required to achieve this change mean that it will come at an enormous cost to our environmental management regime."[58]

On paper, policy analysts observed, South Africa's assemblage of laws met international best practices for mining legislation.[59] Yet leaving aside questions about whom best practices protect, preventive environmental legislation meant nothing without enforcement. That

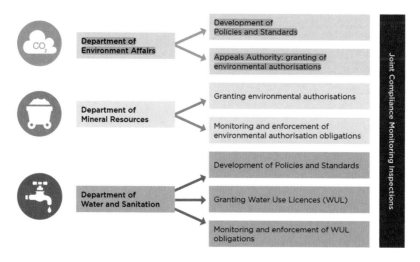

5.15 Chart of the One Environment System established in December 2014. South African Human Rights Commission, "National Hearing on the Underlying Socio-economic Challenges of Mining-Affected Communities."

too required expertise, along with monitoring infrastructures to track emissions and effluents.

Perpetual austerity made it extremely difficult to build robust regulatory capacity. State agencies remained woefully understaffed, qualified inspectors in short supply. One estimate found that the DMR had just ninety-six inspectors to cover 1,757 mining operations in 2016.[60] High vacancy rates made it easy to slip in unqualified political appointees, and backlogs in the permitting process presented ample opportunities for graft.[61] In one survey, mine closure professionals themselves reported that "poorly conducted" baseline studies resulted in inadequate indicators. Air, soil, and water parameters went unmonitored. The licensing process treated each mine as a single entity; no one tracked cumulative impacts. Inadequate funding undermined efforts, which all too often turned into "tickbox exercises."[62] And the global rise of audit culture made tickboxes terribly tempting.[63]

Larger mining companies employed their own closure and remediation experts. But even they sometimes resorted to consultants for EIAs, remediation plans, and compliance monitoring.[64] Mine closure and remediation became big business, complete with professional organizations and trade conferences.[65] Doors continued to revolve as engineers and environmental scientists moved among corporate employers, state agencies, and consulting firms.[66] South African state agencies didn't have the capacity to verify the accuracy of EIAs, which ran 1,500 pages or more; simply reading them consumed more person-hours than agencies had at their disposal.[67]

The task was even more daunting for communities and their advocates. Even for those with mastery of English and a university degree, EIAs constituted a major slog. And that assumed access. State agencies did not make mine closure data readily accessible. Nor did most consultants.[68] One pair of researchers reported waiting nearly two years after filing a data request. When the data finally arrived, they discovered that formatting and content differed by department. "It would appear that not even government has a national overview of mine closure in South Africa," they concluded dismally.[69] No wonder the Centre for Environmental Rights cried foul at shortened timelines.

Democracy for the Capitalists

Struggles over land rights—who has them, what they mean, whether they apply underground as well as on the surface—have long percolated in the background of debates over environmental legislation and the future of mining. Apartheid's algorithm looms especially large in this fight. The Mining Charter (first launched in 2004) is meant to foster socioeconomic empowerment in the mining industry. Stated goals include increasing Black ownership in the mines, more equitable distribution of mining's benefits, and community consultation for new projects. This latter point has proved particularly tricky. What does community consultation mean? Do communities have the right to reject mining outright?

And the real zinger: Who can legitimately speak for a given community?

In the twenty-first century, struggles over these questions play out primarily on the platinum belt. Stretching north of Gauteng across Limpopo and North West provinces, the belt has long supplied most of the world's platinum (six times as much as Russia, the next country down the list). Extraction has been conducted by some of South Africa's largest mine operators, as well as several smaller ones. Large or small, these companies have typically sought access to the subsoil via traditional leaders, claiming that these negotiations fulfill their duty to consult communities.

Postapartheid legislation has reinscribed the chiefly authority conferred by apartheid's Bantustan homelands system.[70] Under the guise of respect for traditional governance, chiefs (often the same ones designated as proxy rulers under apartheid, or their direct heirs) are treated as those with authority to negotiate mining rights. As sociologist Sonwabile Mnwana and others show, however, villagers often dispute the authority of local chiefs to sign mining deals.[71] Chiefs are frequently accused of failing to consult their neighbors and striking deals to enrich only themselves, running roughshod over customary land rights and failing to honor ancestral land purchases. Mining companies, meanwhile, satisfy the Mining Charter's Black Economic Empowerment requirements by seating a handful of powerful players on company boards. By nominally deracializing economic power, the move widens the gulf between elites and the rest.[72]

Current South African president Cyril Ramaphosa offers a prime example of this dynamic. Back in 1982, he had served as the founding leader for the National Union of Mineworkers, South Africa's first formally recognized trade union for Black miners. After 1994, writes Mpofu-Walsh, Ramaphosa became the iconic "Comrade Baron," an unflattering term for "a billionaire who drapes himself in the language of liberation, and . . . a liberation icon who graces the corporate boardroom."[73] An early beneficiary of the ANC's Black Economic Empowerment policy, Ramaphosa's wealth and influence expanded rapidly. He sat on the Lonmin board of directors in 2012, when a massive strike erupted at the company's Marikana mine. As the strike heated up, Ramaphosa, the former union president, infamously emailed fellow board members denouncing the strike's legitimacy and demanding action. The following day, the South African police quelled the strike by gunning down thirty-four workers with assault rifles.

Subsequent hearings determined that Ramaphosa hadn't directly ordered the shootings. But the stain remained. So did the deep disappointment with Comrade Baron. By the 2018 elections, however, the choice was between Ramaphosa and the incumbent president Jacob Zuma, widely seen as having sold the South African state and its assets—including several mines that violated all manner of regulations—to the ultrawealthy Gupta brothers (paradigmatic state capture). Ramaphosa won the election. Mpofu-Walsh's verdict: elite capitalists and liberation leaders collided during apartheid, but "finally merged in the office of the presidency."[74]

The Mining Charter presents itself as a project of preventive governance to remediate apartheid's residues by writing fairer distributions of benefits into the conception of new projects. But companies cling to the lures of residual governance. They take a minimalist approach to the Social Labour Plans mandated by the charter, frequently failing to honor even those paltry promises. Some young villagers, fully cognizant of mining's deadly dangers, don't buy the trade-off between employment and land loss offered by the plans. One young activist summarizes the impossible situation that rural residents find themselves in: "When people are hungry, you cannot convince them that the food they want to eat contains poison. They will just eat and not mind the poison. Truly speaking, most of the youth from my area have no vision at all! If I wanted to work for the mine, they would have employed me long time ago. And then I would work and get sick and die. How can I work for the mines that are here to kill us?"[75] Even small-scale protests over these issues can turn violent, especially when police are summoned. Such conflicts keep the Mining Charter under constant scrutiny, as legislators seek to specify who counted as "interested and affected persons" and what "meaningful consultation" meant.

South African journalist and filmmaker Joseph Oesi spotlights these tensions in his 2016 documentary *Black Lives Matter*. Enraged by the Marikana massacre, Oesi shows how deep injustice continues to plague mining communities in supposedly postapartheid South Africa. "The chief . . . doesn't own the land. The land is ours," explains activist Moekhti Khoda in the film. "Government and mining companies make deals with the chief alone. The municipality, the chief, the mining company, they are one thing. When you fight the municipality, you die; when you fight the mine, you die; when you fight the chief, you die." The Mining Charter enriches only a few, creating a tiny cadre of ultrawealthy "Historically Disadvantaged Persons" and leaving the vast majority impoverished. Khoda doesn't need academic social theory to understand the political ravages of capitalism. Her final verdict indicts the entire democratic dispensation and captures citizen disillusionment: "Democracy is for the capitalists and their children's children."[76]

Screened in June 2016 at a conference organized to discuss the prospect of a People's Mining Charter, the film generated emotional reactions among audience members. Tears welled up as viewers recognized their own struggles on-screen. "We want to see films beyond *Rambo*," said one audience member. "We want to see ourselves on-screen."[77] Nearly three decades after the end of apartheid, opportunities for such recognition remain in short supply. The platinum belt has become a rural reincarnation of the gold belt.

5.16 Oesi's film *Black Lives Matter* screened at the 2016 conference of the Mining Affected Communities United in Action conference. Tara Weinberg.

Dereliction

As struggles over the future continued, mining's past retained its active presence. Their many iterations notwithstanding, neither the Mining Charter nor the spatial development frameworks were tasked with rehabilitating residual land. The six thousand derelict and ownerless mines remained the domain of the Department of Mineral Resources.

By 2017, DMR officials had decided that nearly half of the derelicts did not require any rehabilitation. Another thousand sites were still somewhat operational. Officials had visited but failed to obtain access to another 1,347 sites. That left 753 sites nationwide, which the DMR categorized by rehabilitation urgency. Remediating asbestos mines (estimated at some 2 billion rands) topped the list. Rehabilitating the rest would cost around 46 billion rands, to be split between three government departments. Considering these colossal sums, the number of jobs offered by rehabilitation efforts in 2016–17 was disappointing: a mere 189 "work opportunities" split across four provinces.[78]

The One Environmental System approach introduced in 2014 did nothing to diminish disillusionment with the industry. With remediation in the hands of DMR, civil society remained vigilant. Activists kept the spotlight on the effectiveness of reform, in contrast to its lofty principles. They continued to deploy international examples and experts to bolster their case.[79] Community members also kept up pressure, insistently presenting human faces to government officials.[80] In 2016, such pressure brought matters faced by mining communities to the South African Human Rights Commission (SAHRC), viewed by many as the nation's moral compass.

The commission's hearings elevated the super wicked problem posed by the afterlives of mining to the nation's highest moral stage. Its subsequent report formalized community complaints, giving them epistemic credibility. Commissioners categorically repudiated the DMR's regulatory legitimacy: the DMR was simply "not the appropriate authority for granting and enforcing environmental authorisations," especially since it had proved cavalier enforcing even the easiest measures, such as financial provisioning of remediation funds. Noting that the National Nuclear Regulator lacked human and financial capacity, commissioners directed the state to prioritize funding the NNR to remediate radioactive contamination.[81]

The forceful verdict changed nothing. In March 2021, Mariette Liefferink wrote the commission to inquire about compliance with the 2016 directive. No response. In August, she filed an official request for relevant NNR records, seeking replies to nine formal questions. Question 1: What had the NNR done to comply with the commission's directive? NNR: It had developed a remediation framework and a "proposed funding mechanism," but these hadn't yet reached their final form because the pandemic had paralyzed the process of stakeholder consultation (begging the question of what the agency had been up to in the three-plus years between the SAHRC report's release and the outbreak of COVID-19 in South Africa). Question 2: Status of the NNR's 2015 remediation plan? NNR: "Plans without funds are not implementable." Questions 4, 7, 9: What had been done to protect the public in Kagiso and elsewhere? NNR: "Refer to response in Question 1." Question 8: Tudor Shaft? NNR: Offers to remove Tudor tailings hadn't included rehabilitation, but this would produce "an undesirable situation since members of the public could start occupying an unrehabilitated site" and (see question 2) there was no funding to go further.

As for the rest: more research was required—the standard residual governance playbook.[82] As Liefferink dryly commented to me, "The documents that hold the history of the Wonderfonteinspruit would exceed five metres if stacked on top of each other. The bibliography of relevant literature that has been compiled would, if printed, run to nearly one hundred and twenty pages. The Wonderfonteinspruit is arguably the most complex and most studied catchment in South Africa."[83] The NNR's anemic reply—along with disturbing media reports questioning the Human Rights Commission's independence—prompted Liefferink to compose one of her devastatingly courteous emails to Advocate Jonas Sibanyoni and his SAHRC colleagues in January 2022. "Compliments of the season!" she began cheerfully. (By this point, the commissioners probably knew that didn't bode well.) Liefferink laid out the facts and attached a raft of documentary evidence. The commission had failed to enforce its directive, she concluded, "notwithstanding your commitment on behalf of the SAHRC that non-compliance by an institution 'will be taken to court.'"[84] A month later, still no response.

Meanwhile, scientists at the Council for Scientific and Industrial Research had chugged through that 5-meter stack of research papers and reached similar conclusions. There had been some progress on water treatment and (in one case) wetland restoration. But the CSIR re(re)(re)confirmed that "*there is no universal strategy* for the reclamation and post-mining use of former mining land." Plans had to be site-specific. And some land could never be remediated.[85]

Which took the discussion back to the future. Despite endless remediation failures, CSIR scientists claimed that treating mining as part of a "circular economy" would prevent future disasters. Notwithstanding the long history of failure to meet community needs, they insisted this approach would be "fully inclusive in terms of stakeholder representation." Postmining land use elsewhere included an open-air theater in Sweden, mushroom cultivation in the US, a cathedral in Colombia, and tourism in Romania and Chile. The right incentives would encourage stakeholders to behave responsibly. Mining would become resilient! Green! Sustainable!

In one breath: "The impact on the natural environment is everlasting." In the next: "sustainable mining." Textbook self-devouring growth.[86] Truly, the human capacity for oxymoronic optimism can take your breath away.

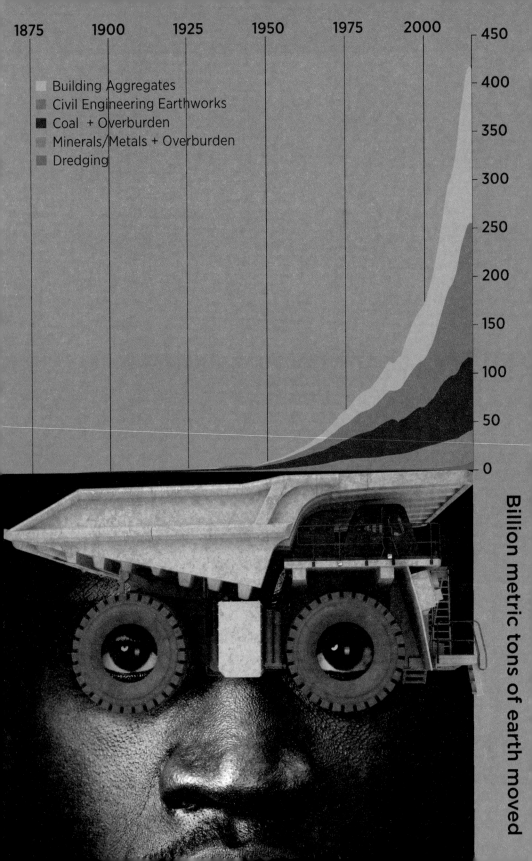

1875 1900 1925 1950 1975 2000

450
400
350
300
250
200
150
100
50
0

Building Aggregates
Civil Engineering Earthworks
Coal + Overburden
Minerals/Metals + Overburden
Dredging

Billion metric tons of earth moved

CONCLUSION
LIVING IN A FUTURE WAY AHEAD OF OUR TIME

THE HUMAN CAPACITY FOR oxymoronic optimism will literally take your breath away if you're among the millions living downwind from the dumps. Actual repair is messy, difficult, and expensive. It's inherently incomplete and utterly unglamorous. Far easier to imagine orderly, well-regulated futures: compliant, funded, absent all corruption. Prosperous, even. Such futures promise to palliate the past. But imagined futures built on "living ruins"—ruins that continue to bleed and cough and seep and surge—cannot heal the wounds etched into lungs and land and lives.[1]

In South Africa, the material and political lives of these living ruins —their technopolitics—are irreducibly entangled with apartheid, its colonial predecessors, and its aftermath. Technopolitics derive their power from their material forms. This makes them difficult to challenge, let alone undo and rebuild. Their endurance is sustained by residual governance of the people and places they've laid to waste.

Residual governance is a prime instrument of the racial contract, an implement for etching inequity into bodies and land. It enables racial capitalism to continue shaping the conditions for "living in [a] future way ahead of our time," as Potšišo Phasha puts it. Whether by design or by default, the spirit and strategies of residual governance have long driven inaction: minimalism, incrementalism, simplification, delay. Manufactured ignorance—whether malevolent or systemic—is a key weapon in the arsenal, one that makes industrial polluters signatories of the racial contract. Inventing a different future begins with identifying and undoing the mechanisms of residual governance that keep the current future in place.

The future in which South Africans live isn't just theirs. Well beyond South Africa, residual governance fundamentally drives Anthropocene accelerations. Study after study shows that the wastes generated by (racial) capitalist production systems continue to expand exponentially. These production systems need residual governance to continue their predation. They need to treat people and land as waste dumps; otherwise, they have trouble turning a profit. And they need that treatment to be legal; otherwise, their shareholders might object. When new regulations threaten, corporations work together to create global entities (like the *Global Acid Rock Drainage Guide*) to define best practices. Often, they find support in guidelines generated by international organizations (like the ICRP), themselves populated by state and industry experts. So when efforts to stave off regulation fail, corporations are ready. They arm their officials with a cobbled-together "international consensus" and send them through revolving doors to help write the rules. When the rules become law, companies hire (or create) consultancies to plan and certify their adherence.

These patterns obtain across a wide range of industries. But let's stick with mining. Just how much waste does extraction generate globally?

Tackling this apparently simple question immediately spotlights a key area of ignorance, because for many places, the relevant data don't exist. Mine waste data do not generate value for shareholders, after all. In fact, they have the opposite effect, highlighting dynamics long externalized by industry. So assessing the worldwide accretion of mineral waste requires combining spotty data with modeling and extrapolation. For example, you can estimate residue volumes by applying "overburden multipliers" to production data ("overburden" being the amount of "sterile" rock requiring removal before getting to the profitable pebbles—the industry's vocabulary is truly wondrous).[2] I conducted this exercise in collaboration with Avery Bick, then a graduate student in environmental engineering. We found that extraction waste, like other Anthropocene trends, has risen exponentially since the second half of the twentieth century.

The graphs depict exponential increases in quantity of earth moved. As I've shown, that's only the beginning of mining's volumetric violence. Acid drainage plagues many abandoned mines. And there are a lot of abandoned mines in the world. Canada has around ten thousand; the US, over half a million.[3] A 2001 UN report guessed that if you included every single shaft and alluvial working, the number of abandoned mines worldwide would run well into the millions.[4] That's not counting mines currently in operation, some of which continue to break records for volume and spread.

A 2017 map of active mines from the *World Atlas of Desertification* could be read as an atlas of future abandonment. Though not everyone would read it that way, of course. When captains of capitalism and their cronies view such atlases, they see growth (*GROWTH*!). And, increasingly, an opportunity for greenwashing. *Look at all the minerals being pulled out of the ground! Soon we'll have even more stuff, and more stuff will help us combat climate change! Fear not: we will grow our way out of the problem. (We just need to find the right consultants.)*

Such interpretations involve a great deal of magical thinking. Recall Charles Mills's argument that racial contract theory reflects lived reality, whereas social contract theory, by ignoring race, can only deal in fanciful abstractions. A similar pattern applies here. The neoliberal intellectuals who serve as capitalism's henchmen insist that limiting growth just isn't realistic.[5] Much like social contract theorists trumpet the universality of European Enlightenment humanism, the neoliberal crowd makes endless growth sound like a matter of global equity. Poor countries have every right to catch up to wealthy nations, they virtuously proclaim, waving away the proposal that wealthy Earthlings reduce their consumption. How unrealistic! Technological innovation will enable all to prosper without sacrifice. For a bunch of self-styled realists, these henchmen are stunningly cavalier with basic physical principles like the conservation of matter. Abstractions serve their purposes

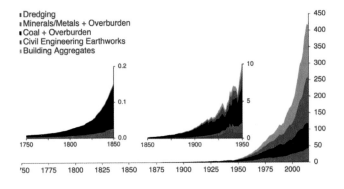

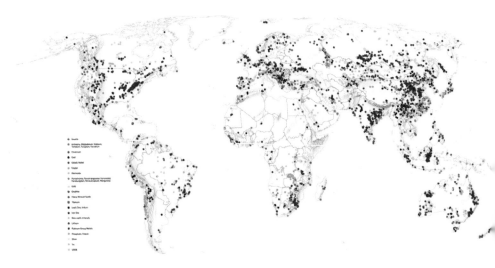

C.1 Exponential growth in earth (ore plus overburden) moved by extraction. Data entry, models, and graphs by Avery Bick.

C.2 Active metal and energy minerals mining sites, 2017. European Commission, "Mining," *World Atlas of Desertification*, April 25, 2019, https://wad.jrc.ec .europa.eu/mining.

far better.[6] Reciting the growth mantra enables them to propound economic theories and policies premised on an infinite planet.

This move doesn't just parallel white supremacy's promotion of color blindness as an Enlightenment ideal—it depends on it. Rhetoric: growth for all. Practice: treat large groups of people and places as disposable.[7] Magical abstractions even enable capitalists to jujitsu moral critiques into support struts. Historian Asif Siddiqi remarks that the "affective nature of critique—people get angry, they emote, etc.— makes the population feel empowered."[8] This conveys the illusion that democracy works, that contrary to activist Moekhti Khoda's accusation, democracy is not just "for the capitalists" (see chapter 5). The magical thinking enabled by the twinned abstractions of race blindness and endless growth enables corporations to embrace climate change as an opportunity rather than a warning. It encourages greenwashing. It promotes carbon accounting as the primary tool for planetary management, distracting from other equally vital aspects of our predicament.

After all, including the full range of issues just complicates matters. Next thing you know, you're looking at a super wicked problem. Those don't have solutions. And the lack of a solution can only lead to despair.

Despair is bad. Hope is good. Keep calm and carry on counting carbon.

A South African vantage point on residual governance also illuminates the interscalar dynamics that drive planetary futures—particularly those that link formal and informal practices.[9] These are most evident in settlement patterns. On one end of the continuum, planned, securitized, gated communities extend upwind of dumps. On the other, unplanned shack settlements perch on or immediately next to dumps. In the middle, residents of formal Reconstruction and Development Programme homes and megaproject communities build informal add-ons and backyard rooms to accommodate family members and paying tenants. These backyard builders buy bricks from informal roadside vendors, who crush cast-off construction materials and mine residues, fashioning these discards into blocks with makeshift ovens. This results in residual homes, built with residual materials, for people treated as residual by the state. The construction materials matter, not just because some of them likely emit radon, but also because the homes they form are unlikely to last long; most large, formal brickmaking factories do not use mine tailings because the presence of sulfates causes bricks to crumble easily.[10] Flimsy materials make for flimsy futures. In these scenarios, time itself becomes residual.

The difference between formal and informal construction materials isn't always clear-cut, mind you. The now-closed African Brick Centre, a factory located just west of Kagiso directly across the road from the highly radioactive Robinson dam, was long rumored to derive up to 20 percent of its feed from nearby uraniferous tailings. I couldn't formally (!) verify this claim. But it wouldn't be unprecedented. Elsewhere in the world, housing, clinics, roads, and recreational spaces have been constructed with materials derived from uranium tailings, resulting in buildings bursting with radon. Grand Junction, Colorado, underwent a Superfund tear-down of such buildings in the 1970s, but radon-emitting houses in Mounana, Gabon, remain inhabited.[11] Examples such as these make clear why both formal and informal settlement residents refuse to separate environment, housing, food, and water when advocating for their rights. When you live and breathe residual governance every minute of every day, then you know it's that very entanglement that harms your body, your life, and your livelihood. You can feel it in your bones.

Reworking (or removing) the dumps also follows the interscalar dynamics of formal and informal labor. Corporations like Mintails acquire dumps to squeeze out the last dregs of profit, then indignantly call for police help when they discover that *zama zamas* have gone to work removing equipment and materials from deserted sites. Meanwhile, *zama zamas* themselves operate at and across different scales. Some are organized into large groups (sometimes even gangs). Others operate peacefully as independent entrepreneurs. Some stick to the surface, combing the piles for scrap metal. Others descend underground, searching for the metal that symbolizes ultimate wealth. Some were salaried mine workers before closures—or personal initiative—drove them into the informal sector. Others have only ever known mine labor as *zama zamas*.

Regulatory mechanisms also combine formal and informal practices. Guidelines and regulations, which can run to thousands of pages, seem the epitome of formality. But most have significant, built-in fuzziness. Recall ALARA, the ICRP's recommendation that operators keep radiation doses "as low as reasonably achievable." ALARA leaves the question of ability in the hands of industry and national regulators. Those in power decide what counts as reason*able* and achiev*able*. These determinations come down to profit (thresholds aren't reasonable if they cut too much into profits) and capacity (enforcement isn't achievable without expert regulators and extensive monitoring). Similarly, standards for mine remediation fall within the limits of practic*ability*, a determination made by the state in close consultation with industrial partners. Notions such as "existing exposure scenarios" somehow manage to formalize wiggle room, again leaving it to industries and their cost-benefit decision trees to set the parameters of "ability."

Finally, knowledge produced by residual governance—and by struggles against its effects—also entangles formal and informal practices. The space between anecdotal evidence (as the knowledge of the desperate and dispossessed is often described) and peer-reviewed institutional

C.3 Robinson Lake and *zama zama* miners (*on the left*), West Rand, 2022. Robinson Lake is one of the most contaminated water bodies in the West Rand. *Zama zama* miners have arrived in this area only recently. Tensions with the community are such that we were strongly advised not to approach; here, we seek to capture the distance with which most people view *zama zama* work. Luka Edwards Hecht.

science is a messy continuum, not a stark chasm. From fundamental science to industry research, from well-funded projects to those conducted under conditions of austerity, from dissident science to citizen science, formal and informal modes of knowledge production often rely on each other.[12]

Capitalist craving for commercial certainty spurs regulatory bureaucracies to distinguish and delimit these knowledge modes, certifying some and dismissing others. Yet bureaucracies, whether in South Africa or elsewhere, can never keep up with the constant creation of contamination. The people discarded by racial capitalism—those who have no choice but to grow crops in contaminated soil and breathe polluted air—serve as frontline sensors. Their bodies are the most sensitive of instruments. But the systems of residual governance are not set up to treat their experiences as data. Instead, these experiences disappear into the tickbox of stakeholder consultation, shoved into quaintness as anecdotal evidence.

The effort to separate formality from informality is always political. When informal practices preserve profits, they receive praise for their practicality and flexibility. When community leaders and activists insist on the epistemic authority of their own bodies to push against the absurdity of purified categories, they face denigration and dismissal.

It's tempting to attribute the increasingly residual character of governance in South Africa to corruption, which has reduced governance to the dregs left after the powerful have drunk their fill. There's certainly plenty of graft to go around, starting at the top. Following a devastating report by South Africa's public prosecutor in late 2016, a commission on state capture (i.e., the capture of the state by oligarchs) was established to investigate further details. Headed by Chief Justice Raymond Zondo, the commission spent over three years hearing testimony and digging into records. The reports rendered in early 2022, totaling some 2,500 pages, present mountains of evidence documenting graft across a staggering range of South African industrial leaders and state officials, far beyond the well-known venal relationship between former president Jacob Zuma and the infamous Gupta brothers. As of this writing, it remains to be seen whether (and which) prosecutions will follow.

There's no question that corruption has undercut progress on mine remediation, among many other things. In South Africa (and elsewhere), corruption at all scales undermines the capacity of already-residual governance. But it would be a grave error to conclude that eliminating corruption would suffice. Doing so would simply repeat the errors of South Africa's Truth and Reconciliation Commission by limiting accountability to individuals, instead of pursuing the hard work of undoing the system upheld by residual governance.

In any case, venality and graft are hardly specific to South Africa (let alone, as mass media often imply, to Africa in general—whatever that means for a continent with fifty-four countries, whose total landmass exceeds that of the US, China, and India combined). Many of the world's dominant nations have legalized profiting from politics.[13] At the global scale, offshore tax havens hide profits and bribes, while dumping licenses hide garbage. These forms of legalized corruption buttress residual governance and the modern forms of racial capitalism that it propagates. Of course eliminating corruption would help. Imagine! But this alone would not be enough to reverse the inequities and contaminations that compose the Anthropocene.

South Africa's liberation struggle presented real opportunities to turn back the tide of racial capitalism, and of the racial contract upon which it rests.[14] Or, failing that, to significantly dampen its worst excesses. And it's important to recognize that change has happened. It matters that racial segregation and differential treatment are no longer legal, and that all adults can vote. Millions of people have received homes that transformed their lives. Millions receive state payments that can make the difference between eating and starving. Nevertheless, racial capitalism persists, lodged in infrastructure and sustained by residual governance. The liberation struggle is far from over. But like apartheid, it's taken a new form.

The people who populate this book show what that struggle looks like now, what's required to invent a different future. Resilience and hope are not abstract affective conditions. They're work. As community leaders, activists, scientists, lawyers, artists, and journalists know all too well, this work is difficult and frustrating. It requires study after study, hearing after hearing, for decades. Reports and testimonies are necessary (if insufficient) tactics in renewing the struggle against infrastructural inequity. The repetitions themselves index the power and persistence of a racial capitalism so deeply embedded in infrastructure that it stays strong through the very political upheavals that aim to topple it.

Capitalists and the states they capture prefer generalizable solutions. Solutions whose design and transaction costs can be minimized by economies of scale. (Water treatment plants!) Solutions that can be grandly proclaimed from a political podium. (Megaprojects!) Solutions that will make shareholders happy. (Remine the dumps!) Solutions whose parameters will, thanks to some revolving-door action, limit what regulations can demand. (Existing exposure scenarios!) Solutions that oversimplify wicked problems. Solutions that address part of a problem while declaring the rest beyond the solutioneer's jurisdiction. Solutions with certainty. Walk-away solutions. Final solutions, though no one dares to use that phrase anymore.

Struggles against residual governance challenge these fantasies. They shine a spotlight on human and molecular details to highlight the flaws of simplified solutionism. They insist that details matter, and not just on a small scale. Details matter as evidence in lawsuits, parliamentary hearings, and public inquiries. Presented as data, photographs, or personal stories, they can create urgency. Details enable community activists to tack between specific forms of pollution and broader trends in order to maximize their leverage. In a city-region as fragmented as Gauteng's, and a nation-state that seems to spend more energy eating itself alive than serving its citizens, blanket policies mean little if they can't be enforced at the microlocal level. This makes specifics—down to the placement and reliability of pollution sensors—all important. Like all liberation struggles, fights against residual governance begin as local struggles.[15]

Local action is vital for holding governments, corporations, and their infrastructures to account. Yet identifying the devils in the details—those that belie the promises of the powerful—is not enough to counter the technopolitics of the racial contract, whose expression in infrastructures too often remains invisible (at least to the beneficiaries, including many who aren't active signatories). Among many other things, fights against residual governance are fights for recognition. Recognition of personhood, for starters—and of how infrastructural inequalities deny personhood.

This liberation struggle demands recognition that political and economic justice can only be achieved via infrastructural justice. It insists that infrastructures be defined not only by their productive potential, but also by their detritus, their damage, and their destructive indifference. For those who fight against it, the opposite of residual governance is not a souped-up technocratic machine. It's the creation and maintenance of systems and infrastructures that not only recognize and respect their full personhood (their voices, their bodies, their aspirations), but also have mechanisms for sustaining that recognition and respect over time.

Absent a wholesale upheaval of racial capitalism, securing any form of remediation requires building a case place by place. Officials and experts can successfully invoke international expertise to shore up their policies, but when communities try similar tactics (for example, by citing international studies that link specific contaminants to cancer), the response is almost invariably something like *But what about here?* Or *Careful, exposure doesn't mean uptake!* Or *Prove it!* International solidarity is vital in the struggle against residual governance. But staggering resource imbalances make building the necessary networks difficult. Leveraging them is even harder.

Nevertheless, people continue to chip away at the challenges. They have no real choice. Their lives are at stake. After years spent waiting for the state to come to their rescue, they know that the lifeboats are too few, too small, too leaky, and too easily snatched away. Millions of people feel abandoned alongside the mine sites that surround them. Millions feel treated like waste by men who once promised liberation, and now profit from it. And not just in South Africa.

The abandoned cannot afford to think in boxes. Not for them the neat divisions between environment and society, radiotoxicity and chemical toxicity, surface and subsurface rights, scientific evidence and lived experience. They know better. Their bodies and homes form the front lines where these elements entangle and collide with other residues of racial capitalism. They certainly understand the importance of strategic compartmentalization. If designating a zone "nuclear" improves the odds of obtaining a long-promised home, then radiation measurements become a useful technopolitical tool. But for the abandoned, the first goal is not an abstractly imagined pristine environment. It is, quite simply, a breathable, drinkable, edible one, sheltered from the worst molecules of racial capitalism and the Anthropocene. Most other goals require achieving this one first.

The abandoned and the vulnerable know how the world will look if it remains mired in residual governance. Such a world will be inside-out, hollow. In it, almost everyone is a *zama zama*, digging through the toxic dregs. Some will work in gangs. Some will go it alone. Some will form community; they'll be better off than others. Most will struggle to breathe.

The alternative? No retreat. No surrender.

ACKNOWLEDG-MENTS

I first flew over the Witwatersrand plateau in 1998, four years after South Africa's first democratic elections. Back then, I didn't understand what I was seeing. That began to change in 2003–4, when I spent over a year in South Africa researching the history of uranium mining for *Being Nuclear*. I decided to save the topic of mine waste for postbook essays. Eventually I realized that I had another whole book on my hands.

During that time, it has gradually dawned on me that my personal story intersects with some of the histories told here—and that this bears acknowledging. My US citizenship is the unplanned flotsam of waves of American imperialism. When the Dominican Civil War broke out in 1965, US president Lyndon B. Johnson sent marines to quash the revolution and prevent the island from becoming another Cuba. My pregnant Dominican mother and Swiss father were living in Santo Domingo then, but my father's Swissness qualified them for evacuation. They were shipped to Puerto Rico, where they stayed until I emerged as the only US citizen in the family. To make me into an American, my parents decided to speak English at home (albeit heavily peppered with Spanish) and sent me to American schools as we moved around Latin America and Europe. Peer pressure flattened my Hispano-Germanic accent into a white American movie inflection. In 1982, I moved to the continental US to attend college—another family first.

Only much later did I come to see how the mechanisms of racial capitalism had played a significant role in this story. In 1951, my father escaped a difficult home in Switzerland by moving to postwar London. He spoke no English and had only a high school education, but he did have three other languages, a knack for arithmetic, and a jovial demeanor. The Royal Bank of Canada (RBC) hired him as a mailroom clerk. He learned English on the fly, spending a couple of years in a dismal boarding house, eating sardines for breakfast and feeding the coin-operated heater at 2 a.m. (This very design would become a prototype for water and electric metering in poor households in postapartheid South Africa, and a target of service delivery protests, as Antina von Schnitzler has shown.) Upon discovering his language skills and winning personality, RBC promoted my father to their expatriate division. He jumped on the first occasion to move to the Caribbean, landing first in then-British Guyana before moving to the Dominican Republic in the early 1960s. It took Peter James Hudson's analysis of how RBC and other banks actively undermined Caribbean sovereignty in the early twentieth century for me to realize that my father had participated in a later phase of that same exploitation. A bit player to be sure, just a junior account manager among many . . . but an expat nonetheless, privileged enough to qualify for evacuation when the American marines arrived to quell incipient revolution.

My father's expat status enabled us to live beyond the means afforded by his salary alone. Benefits for RBC expats included rent, private schooling, and regular "home leave." As Charles Mills would say, we may not have signed the racial contract, but we definitely benefited from it. The main trade-off was that we had to move every three years or so. That's how I found myself in France in 1976, the year of the Soweto uprising.

I was just a year younger than Hector Pietersen when he was murdered by apartheid police. Perhaps that's why my parents shielded me from the news, why my early political memories include Watergate and the death of Spanish dictator Francisco Franco (I'd never seen my father so excited about anything), but not the depredations of apartheid in South Africa. The comfortable cocoon of our whiteness certainly played a role in that silence. That probably also explained why my parents never commented on my paradoxical friend group: a Black American girl named Marguerite and a white South African girl named Michelle, both daughters of diplomats posted to Paris. Photos from 1976 and 1977 show us playing in the garden, both of them towering over puny little me, all of us giggling uncontrollably. I still wonder what conversations Marguerite's family had about this arrangement. Mine was mute, a reticence born of privilege that pretended color blindness. I'm guessing Michelle's was too. Whatever the case, I learned very little about apartheid before I started college.

By the time I graduated, I'd joined the ever-louder calls of the anti-apartheid movement for universities and corporations to divest their South African holdings. I finally understood just how strange my simultaneous friendship with Marguerite and Michelle had been. But it wasn't until my first term of grad school, when I applied to the National Science Foundation for funding, that I realized that I myself wasn't fully white by US demographic categories.

I remember the exact moment this happened. Joyce, our African American department secretary, was kind enough to let grad students use her typewriter after hours to fill in application forms. It was an advanced electric model, so your fingers didn't get trapped between the keys; this was 1986, after all. We all converged in Joyce's office to take turns and compare notes. When my turn came, I rolled the form up along the platen to reach the demographic section. The top choice most emphatically did not apply to me: "White, not of Hispanic origin." So, Hispanic it was. (I don't remember an option to check more than one box back then.) Another student peered at my form and dropped a bitter comment about my gaming the system. I felt genuinely confused. This was a government document, after all; I had to sign a statement swearing I'd told the truth, and the truth was that I was of Hispanic origin. Four years in the continental US hadn't yet taught me why it mattered that neither my name nor my skin tone corresponded to what Americans thought of as Hispanic (though I'd already gotten tired of explaining to mainlanders that Puerto Rico was, indeed, part of the United States). I learned an important lesson in that moment of confusion. That's also when I finally began to think seriously about race in America. Way too slowly, I grew to see deep connections between South African apartheid, the depth and horror of American racial politics, and my own peculiar origin story. All of which informs my approach to this book.

It still feels miraculous that my friends and colleagues in South Africa have tolerated my bumbling presence for the last two decades. This book wouldn't exist without them. Nor would it exist without the special community that emerged at the intersection of science and technology studies and African studies in the early 2000s and 2010s at the University of Michigan, or without the multiyear grant awarded by the Mellon Foundation to the University of the Witwatersrand and the University of Michigan to foster a rethinking of the humanities in Africa. I have benefited from the collegiality and generosity of these entangled communities in ways impossible to detail fully. Here's a start.

Keith Breckenridge and Catherine Burns have long provided sustenance of all sorts during my repeated trips, not to mention sharp analysis and merciless commentary. Pamila Gupta was a wonderful thinking, writing, and editorial partner for our series "Toxicity, Waste, and Detritus." Tara Weinberg was an intrepid research assistant, as well as a mighty scholar in her own right. The comments of Derek Peterson, ever a sparkling sparring partner, helped me rework the argument at a crucial juncture. Anne Berg taught me the real meaning of trash talk and reminded me that scholarship benefits from howling laughter. Hannah le Roux partnered with me on a search for radioactive building materials, which led to our coauthored essay "Bad Earth"; she also offered crucial feedback on the draft manuscript. Lesley Green's amazing and generous review sharpened the manuscript, as did her scholarship. Toward the end of this project, I had the tremendous good fortune of meeting Potšišo Phasha, who not only gave permission to include some of his work but also proved a stunning interlocutor for my own; I'm still thinking about some of his comments on the penultimate draft. Chaz Maviyane-Davies patiently and brilliantly rendered the essence of the argument in visual form.

Anyone who's taken one of Mariette Liefferink's toxic tours knows she's an endless font of information and generosity. She provided access to thousands of primary source documents and commented on an earlier draft of this manuscript. Nkosinathi Sithole took considerable time from a busy schedule to help me understand some of the legal ramifications of the events I was tracking. Jeffrey Ramoruti toured me and Tara around Tudor Shaft and shared his experiences and documents. Joseph Oesi shared his film work and insights. Frank Winde walked me through some of the scientific research I investigated. Kim Kearfott lent me a Geiger counter and offered essential perspective from her own experience. Neil Overy helped me understand the Mining Charter.

Farther afield, Itty Abraham, Ken Alder, Soraya Boudia, and Peter Redfield have been sparkling interlocutors for decades, always challenging me to think further and deeper, always with friendship. Others who offered essential collegiality, insight, and forums in which to workshop my arguments include Omolade Adunbi, Mohammed Rafi Arefin, Kelly Askew, Richard Ballard, Faeeza Ballim, Javiera Barandiarán, Andrew Barry, Philippe Behra, Ele Carpenter, John Carson, Nick Caverly, Simukai Chigudu, Shadreck Chirikure, Lane Clark, Kristen Connor, Kenny Cupers, Robyn D'Avignon, Jacob Dlamini, Kevin Donovan, Jatin Dua, Malcom Ferdinand, Rosalind Fredricks, Claudia Gastrow, Sally Gaule,

Zsuzsa Gille, Josh Grace, Joel Howell, Tracy-Lynn Humby, Fredrik Albritton Jonsson, Dolly Jørgensen, Finn Arne Jørgensen, Abby Kinchy, Jonathan Klaaren, Scott Knowles, Silvia Lindtner, Julie Livingston, Marie Ghis Malfilatre, Daniel Masekameni, Andrew Mathews, Nayanika Mathur, Chakanetsa Mavhunga, Achille Mbembe, Stephan Miescher, Gregg Mitman, Lumkile Mondi, Sophie Sapp Moore, Ngaka Mosiane, Lisa Nakamura, Tholithemba Ndaba, Rafe Neis, Sarah Nuttall, Ruth Oldenziel, Davide Orsini, Buhm Soon Park, Emma Park, Roopali Phadke, Kavita Philip, Anne Pitcher, Ellen Poteet, Amanda Power, Markus Reichardt, Tasha Rijke-Epstein, Liz Roberts, Richard Rottenburg, Nafisa Essop Sheik, Sheila Sheikh, Asif Siddiqi, Johanna Siméant, Jessica Smith, Stephen Sparks, Alex Stern, Lynn Thomas, Megan Vaughan, and Daniel Williford. Edward Jones-Imhotep and Sonita Moss grounded me during especially difficult times.

A fellowship from the National Endowment for the Humanities enabled me to complete the manuscript, while funds from Stanford helped to pay for research. Generous subventions from the History Department, the Freeman Spogli Institute, and the School of Humanities and Sciences enabled the inclusion of artwork as well as the book's Open Access status; special thanks to Mike McFaul, Caroline Winterer, and the H&S deans for their support. Richard Roberts introduced me to African studies over three decades ago and has remained a staunch supporter ever since.

Kevin Chen kept me sane, chased down references, facilitated copyright permissions, and so much more; for two years, he was and did everything. Rebecca Wall and Katja Schwaller helped with desktop research and photo permissions; Ayanda Mahlaba with translation; Avery Bick with modeling the inside-out Earth; Adele Stock with the king of the dogs. Scott Fendorf and Juan Pacheco analyzed dirt samples. Others at Stanford who nourished this project include Joel Cabrita, Jim Campbell, Dean Chahim, Esther Conrad, J. P. Daughton, Sibyl Diver, Aliyah Dunn-Salahuddin, Paulla Ebron, Rod Ewing, Jim Ferguson, Zephyr Frank, Trevor Getz, Kyle Harmse, Laura Hubbard, Herb Lin, Steve Luby, Ana Minian, Alesia Montgomery, Megan Palmer, Grant Parker, Emily Polk, Steven Press, Jasmine Reid, Harold Trinkunas, Fred Turner, Jun Uchida, and Vannessa Velez. I'd be flailing around desperately without Maria Van Buiten, Kai Dowding, Colin Hamill, Burçak Keskin Kozat, John Lee, Maria Moreno-Lane, and Art Palmon. Anyone who's had the privilege of working with Brenda Finkel or Tracy Hines knows that they're in a class of their own.

Collaborating with the Bayview Hunters Point Community Advocates (BVHPCA) has shaped my perspective on environmental justice—and on the city that I've come to call home—in ways that no amount of reading could. To Dalila Adofo, Tony Kelly, Michelle Pierce, Karen Pierce, and members of the BVHPCA board: my infinite thanks for the warm welcome you've offered me and my students, and for all that you have taught us.

Learning to write takes a lifetime. Those from whom I learned the most about this craft will forever have my gratitude: Michael Brenson in high school; Jay Slagle since college; Nina Lerman in grad school; Paul Edwards since our fight about discourse in 1992; and Sally Davies just a few years ago. From our first conversation to the moment of book submission, Elizabeth Ault at Duke University Press has been the most stellar (and fun!) of editors. Ben Kossak cheerfully guided me through the many details of the submission process; bless him and his patience. Liz Smith guided the book through its final phases; Karen Fisher offered perfectly pitched copyediting; Matthew Tauch designed the interior with talent and patience; and Celia Braves handled the index.

When you grow up as an only child whose cultural referents and political commitments differ dramatically from those of your parents, chosen family is everything, whatever distances introduce themselves over the course of a lifetime. Ginny, Ivan, Jay, Karin, Nina: that's been you for most of my life now. To my Ann Arbor stars: Alex and Terri, Anne, Doug, Kali, Marlyse and Roger, Susan and T. R. And to those who have also become family by legal or biological definitions: Lauren, Maya, Elaine, and Don, and the Edwards clan.

As for the family that puts up with me on a daily basis . . . I can't imagine writing a book without Paul at my side: partner in all things, ruthless editor, best friend. Our offspring Luka grew up with daily rants about apocalypse and injustice; they cheerfully photographed the mine dumps on my last research trip and constantly teach me to see the world differently.

Finally, there's the family who make South Africa a spot of home, whose kinship began in the most unexpected of ways, and whose love and laughter keep it all real. Nolizwi, this book is for you and the family you've joined to mine: Cyprian, Ayanda, Azile, Kwakhanya, Salizwa, Sneh, and the children still to come.

NOTES

INTRODUCTION

1   *Sithole* is pronounced "See-TOH-lay." This story is adapted from Nieftagodien and Gaule, *Orlando West, Soweto*.

2   Quoted in "The June 16 Soweto Youth Uprising."

3   Editorial by the African Teachers Association of South Africa (Atasa) in *The
World*, January 6, 1975, quoted in Nieftagodien and Gaule, *Orlando West,
Soweto*, 69–70.

4   For introductions to the history of the Soweto uprising and subsequent political and historiographical debates about that history, see Ndlovu, *The
Soweto Uprisings*; Nieftagodien, *The Soweto Uprising*.

5   Mrs. Sithole's account quoted in Nieftagodien and Gaule, *Orlando West,
Soweto*, 75.

6   United Nations Security Council Resolution 392 (1976) on killings and violence by the South African apartheid regime in Soweto and other areas, adopted by the Security Council at its 1930th meeting, June 19, 1976, https://
digitallibrary.un.org/record/93718?ln=en.

7   The historiography on the Soweto uprising has long argued that "riot" was
a racist characterization by the apartheid government. Elizabeth Hinton
makes a similar point about so-called race riots in the United States, showing how the idea of a riot was used to justify militarized police response to
Black insurgency. See Hinton, *America on Fire*.

8   Cole, *House of Bondage*, 20.

9   Knape, "Notes on the Life of Ernest Cole," 22–23.

10   Woodson, *The Mis-education of the Negro*, 4.

11   Bhimull et al., "Systemic and Epistemic Racism."

12   Cole, *House of Bondage*, 21.

13   Mills, *The Racial Contract*, 43.

14   I first developed this meaning of technopolitics in Hecht, *The Radiance of
France*. For a deeper discussion of technopolitics in apartheid South Africa,
see Edwards and Hecht, "History and the Technopolitics of Identity"; von
Schnitzler, *Democracy's Infrastructure*.

15   Mills, *The Racial Contract*, 43, 11.

16   Edwards and Hecht, "History and the Technopolitics of Identity"; Edwards, "The Mechanics of Invisibility."

17   Hecht, *Being Nuclear*, appendix.

18   My analysis of the evidence is influenced by a wide range of scholarly fields, including STS (science and technology studies, my field of origin), African studies, history, anthropology, environmental humanities, Anthropocene studies, visual culture, scholarship on racial capitalism and extractivism, critical race theory, and more. Readers will find some of this in the notes, but I have not tried to capture the full influence of all fields on my thinking.

19   Kojola and Pellow, "New Directions in Environmental Justice Studies."

20   Iheka, *African Ecomedia*.

21   Maviyane-Davies, *A World of Questions*.

22   Legassick and Hemson, "Foreign Investment."

23   Quote from Mills, *The Racial Contract*, 26–27.

24   N. Alexander, *One Azania, One Nation*; Legassick and Hemson, "Foreign Investment"; N. Alexander, *An Ordinary Country*.

25   Neville Alexander, "Nation and Ethnicity in South Africa," quoted in "Racial Capitalism, Black Liberation, and South Africa," *Black Agenda Review*, December 16, 2020, https://www.blackagendareport.com/racial-capitalism -black-liberation-and-south-africa.

26   Legassick and Hemson, "Foreign Investment."

27   N. Alexander, "Nation and Ethnicity in South Africa."

28   Gilmore, *Golden Gulag*.

29   Vergès, "Capitalocene, Waste, Race, and Gender."

30   Mbembe, *Critique of Black Reason*, 136–37.

31   Hudson, *Bankers and Empire*; Jenkins, *The Bonds of Inequality*; Jenkins and Leroy, *Histories of Racial Capitalism*.

32   Ferdinand, *Une écologie décoloniale*; Ebron, "Slave Ships Were Incubators for Infectious Diseases."

33   Thomas Sankara quoted in Ferdinand, *Une écologie décoloniale*, 35.

34   Ferdinand, *Une écologie décoloniale*, 16, 54–55, 253.

35   Haraway, *Staying with the Trouble*; J. W. Moore, *Capitalism in the Web of Life*; Tsing et al., *Feral Atlas*.

36   Hecht, "Interscalar Vehicles."

37   Asafu-Adjaye et al., "An Ecomodernist Manifesto."

38   Vergès, "Capitalocene, Waste, Race, and Gender."

39   Zimring, *Clean and White*; Pulido, "Flint"; Millar, "Garbage as Racialization"; Reno and Halvorson, "Waste and Whiteness"; Krupar, "Brownfields as Waste"; Dillon, "Race, Waste, and Space."

40   Livingston, *Self-Devouring Growth*.

41   Winde, "Uranium Pollution of Water," 44.

42   Among many others: Bullard, *Unequal Protection*; Bullard and Wright, *The Wrong Complexion for Protection*; Pellow, *Resisting Global Toxics*; Pellow

and Brulle, *Power, Justice, and the Environment*; Taylor, *The Environment and the People*.

43   Mills, "Black Trash," 89. This quote preserves the capitalization practice of the original.

44   Green, *Rock/Water/Life*, 206.

45   Dlamini, *Safari Nation*.

46   Mavhunga, *The Mobile Workshop*; Mavhunga, *What Do Science, Technology, and Innovation Mean from Africa?*; Mavhunga, *Transient Workspaces*.

47   Finney, *Black Faces, White Spaces*; Taylor, *The Rise of the American Conservation Movement*.

48   McDonald, *Environmental Justice in South Africa*.

49   Von Schnitzler, *Democracy's Infrastructure*; Redfield and Robins, "An Index of Waste."

50   Humby, "The Spectre of Perpetuity," 108.

51   Hepler-Smith, "Molecular Bureaucracy."

52   Green, *Rock/Water/Life*; Dubow, *Scientific Racism*; Dubow, *A Commonwealth of Knowledge*; Beinart and Dubow, *The Scientific Imagination in South Africa*; Hecht, *Being Nuclear*; McCulloch, *South Africa's Gold Mines*; McCulloch, *Asbestos Blues*; McCulloch, "Asbestos, Lies and the State"; Packard, *White Plague, Black Labor*; Braun, *Breathing Race into the Machine*.

53   Dugard and Alcaro, "A Rights-Based Examination."

54   Mills, *The Racial Contract*, 18–19. Black intellectuals working in other scholarly traditions have reached compatible conclusions, of course, including Fanon, *Wretched of the Earth*; Mbembe, *Critique of Black Reason*.

55   Proctor and Schiebinger, *Agnotology*; Oreskes and Conway, *Merchants of Doubt*; Hecht, *Being Nuclear*.

56   Mills, *The Racial Contract*, 97.

57   Liboiron, Tironi, and Calvillo, "Toxic Politics," 333.

58   Haraway, *Staying with the Trouble*.

59   Mbembe, *Out of the Black Night*, 10–12. I'm taking a slight liberty here: Mbembe doesn't confine himself to South Africa, writing "that there's no better laboratory than *Africa* to gauge" those limits (emphasis mine).

ONE YOU CAN SEE APARTHEID FROM SPACE

1   Bobbins and Trangoš, *Mining Landscapes of the Gauteng City-Region*; Trangos and Bobbins, "Gold Mining Exploits."

2   The description of *zama zama* experience is based on a growing body of literature about them, including Thornton, "Zamazama"; Lewis and Zack, *Undercity*; Morris, "Shadow and Impress."

3   Morris, "Death and the Miner."

4   Rittel and Webber, "Dilemmas in a General Theory of Planning," 156.

5   Nelson, *Body and Soul*.

6   Levin et al., "Overcoming the Tragedy of Super Wicked Problems," 124.

7   Hecht, "Interscalar Vehicles."

8   Yusoff, *A Billion Black Anthropocenes*.

9   Hickel et al., "Imperialist Appropriation."

10   Global Justice Now, "Honest Accounts 2017."

11   Rodney, *How Europe Underdeveloped Africa*.

12   Winde, "Uranium Pollution of Water," 27.

13   Boudia et al., *Residues*. Our approaches resonate and overlap, as do some of our terms. My own early thinking on the question of residues appears in Hecht, "Residue"; Hecht, "The African Anthropocene"; Hecht, "Interscalar Vehicles"; Hecht, "Human Crap."

14   Beckett and Keeling, "Rethinking Remediation."

15   Mudd, "Global Trends in Gold Mining."

16   Pigou, *The Economics of Welfare*; Coase, "The Problem of Social Cost."

17   Knowles, "Slow Disaster in the Anthropocene"; Knowles, *The Disaster Experts*.

18   Parvin and Pollock, "Unintended by Design."

19   Fressoz, *L'apocalypse Joyeuse*.

20   A foundational text: Bauman, *Wasted Lives*. Subsequent literature is too extensive to cite; one example is Gidwani and Reddy, "The Afterlives of 'Waste.'" There is also an extensive literature on capitalism and settler co-lonialism's treatment of space—especially Native American and First Na-tion lands—as wasteland; for an example from uranium mining, see Voyles, *Wastelanding*.

21   Bullard, *Unequal Protection*; Bullard and Wright, *The Wrong Complexion for Protection*; Pellow, *Resisting Global Toxics*; Pellow and Brulle, *Power, Justice, and the Environment*; Taylor, *The Environment and the People*; Fortun, *Advocacy after Bhopal*.

22   Mendes, "Molecular Colonialism." See also Liboiron, *Pollution Is Colonialism*.

23   Livingston, *Debility and the Moral Imagination*; Ledwaba and Sadiki, *Broke and Broken*.

24   Guinier and Torres, *The Miner's Canary*.

25   Abrahams, "Fancies Idle."

26   Schrader and Winde, "Unearthing a Hidden Treasure."

27   Hecht, "Interscalar Vehicles."

28   Breckenridge, "Made in America."

29   Reynard et al., "What a Bone Arrowhead from South Africa Reveals."

30   Lombard, *On the Minds of Bow Hunters*.

31   Chirikure, "Metals in Society"; Chirikure, *Metals in Past Societies*.

32   Quoted in Thain, "Ancient Gold-Mining Sites," 243.

33   Friede, "Iron Age Mining in the Transvaal." See also Kuklick, "Contested Monuments."

34 Adas, *Machines as the Measure of Men*.

35 Miller et al., "Indigenous Gold Mining in Southern Africa," 92.

36 Bench Marks, "'Waiting to Inhale,'" 46.

37 Dlamini, *Safari Nation*, 75.

38 Bench Marks, "'Waiting to Inhale,'" 57.

39 Bench Marks, "'Waiting to Inhale,'" 30–31.

40 Gevisser, *Lost and Found in Johannesburg*, 46.

41 Mpofu-Walsh, *The New Apartheid*.

42 Breckenridge, "'Money with Dignity.'"

43 Hudson, *Bankers and Empire*.

44 W. R. Johnson, "Getting Over by Reaching Out."

45 Wendel, "Radioactivity in Mines and Mine Water."

46 Hecht, *Being Nuclear*.

47 Edwards, "Entangled Histories."

48 Zalasiewicz et al., "When Did the Anthropocene Begin?"

49 Sheree Bega, "Living in SA's Own Chernobyl," *Saturday Star*, January 8, 2011.

50 Hecht, *Being Nuclear*, chaps. 6 and 8.

51 Fressoz, *L'apocalypse joyeuse*.

TWO THE HOLLOW RAND

1 Turton et al., "Gold, Scorched Earth and Water."

2 Bench Marks, "'Waiting to Inhale,'" 15.

3 Stoch, Winde, and Erasmus, "Karst, Mining and Conflict."

4 Guy and Thabane, "Technology, Ethnicity and Ideology."

5 Sello, "Rivers That Become Reservoirs"; Hoag, "'Water Is a Gift That Destroys.'"

6 Du Toit, "Background Report on Communities at Risk."

7 Weizman, *Hollow Land*.

8 Stoch, Winde, and Erasmus, "Karst, Mining and Conflict."

9 Bremner, "The Political Life of Rising Acid Mine Water."

10 Ariza, "River of Vitriol."

11 Agricola, *De re metallica*, 6.

12 Agricola, *De re metallica*, 8.

13 Agricola, *De re metallica*, 14.

14 Parvin and Pollock, "Unintended by Design."

15 Edwards and Hecht, "History and the Technopolitics of Identity."

16 Reichardt, "The Wasted Years," 66.

17 Quoted in Reichardt, "The Wasted Years," 77.

18 Frost, "Report on an Investigation into Mine Effluents."

19 Stoch, Winde, and Erasmus, "Karst, Mining and Conflict."

20 Winde, "Uranium Pollution of Water," 26.

21   Stoch, "Krugersdorp Game Reserve."

22   Hecht, *Being Nuclear*, chap. 6.

23   Stoch, "Some Controversial Health Related Issues," 7.

24   Jordaan et al., "Final Report of the Interdepartmental Committee on Dolomitic Mine Water"; Pulles, "Radionuclides in South African Gold Mining."

25   Oreskes and Conway, *Merchants of Doubt*; Proctor and Schiebinger, *Agnotology*; Proctor, *Golden Holocaust*.

26   Adler et al., "Water, Mining and Waste."

27   Auf der Heyde, "South African Nuclear Policy since 1993."

28   "Draft: South African Energy Policy: Discussion Document: Comment," October 2, 1995, 1, 11, Chamber of Mines archives.

29   A. H. Munro to M. Golding, March 2, 1995, Chamber of Mines archives.

30   Murphy, *The Economization of Life*.

31   Munro to Golding, March 2, 1995.

32   D. G. Wymer, "Note for the Record: Meeting between the Chamber and Marcel Golding, Cape Town, 7 June 1995," June 14, 1995, Chamber of Mines archives.

33   Winde, "Uranium Pollution of Water"; IWQS, "Report on the Radioactivity Monitoring Programme."

34   IWQS, "Report on the Radioactivity Monitoring Programme," 13.

35   IWQS, "Report on the Radioactivity Monitoring Programme," 17.

36   IWQS, "Report on the Radioactivity Monitoring Programme," 14.

37   "Ideal" meant "suitable for lifetime use"; "acceptable" meant "suitable for use, rare instances of negative effects." IWQS, "Report on the Radioactivity Monitoring Programme," 18.

38   IWQS, "Report on the Radioactivity Monitoring Programme," 30.

39   IWQS, "Report on the Radioactivity Monitoring Programme," 32.

40   Wade et al., "Tier 1 Risk Assessment of Selected Radionuclides," vi.

41   IWQS, "Report on the Radioactivity Monitoring Programme"; Wade et al., "Tier 1 Risk Assessment of Selected Radionuclides"; Coetzee et al., "Uranium and Heavy Metals in Sediments"; Coetzee, Winde, and Wade, "An Assessment of Sources."

42   Wade et al., "Tier 1 Risk Assessment of Selected Radionuclides," 53.

43   Wade et al., "Tier 1 Risk Assessment of Selected Radionuclides," 56.

44   Marda Kardas-Nelson, "Rising Water, Rising Fear: SA's Mining Legacy," *Mail and Guardian*, November 12, 2010.

45   Nelson Mandela, "No Water, No Future," Nelson Rolihlahla Mandela, August 28, 2002, http://www.mandela.gov.za/mandela_speeches/2002/020828 _waterdome.htm.

46   Quoted in Death, "Troubles at the Top," 560.

47   Death, "Troubles at the Top."

48   "Mariette Has 'Em Shell-Shocked," *News24*, August 22, 2002, https:// www.news24.com/News24/Mariette-has-em-shell-shocked-20020822.

49   Funke, Nienaber, and Gioia, "Interest Group at Work."

50 Quotes from "Our Tainted Lives in Mining's Ground Zero," *Cape Argus*, April 12, 2008; Sheree Bega, "Living in Fear of a Toxic Tsunami; Far West Rand Residents Claim Poisoning," *Star*, April 12, 2008. There are dozens more articles.

51 International Network for Acid Prevention, "GARD Guide."

52 Remaining attendees came from a smattering of European and central Asian countries, along with a handful of experts from the United States and China. Merkel, Planer-Friedrich, and Wolkersdorfer, *Uranium in the Aquatic Environment*.

53 Steffen Flath, foreword to Merkel, Planer-Friedrich, and Wolkersdorfer, *Uranium in the Aquatic Environment*, xix.

54 Beckett and Keeling, "Rethinking Remediation"; Voyles, *Wastelanding*; Winde, "Uranium Pollution of Water."

55 Peter Waggitt, interview with author, Darwin, Australia, August 2001.

56 Waggitt, "Changing Standards and Continuous Improvement," 650–51. Such an approach, he noted, required state backing: "Few organisations other than central governments are likely to have the capability to provide adequate resources to manage the situation effectively and to the degree expected by the community."

57 McGill, Fox, and Hughes, "Rehabilitation of the Nabarlek Uranium Mine," 736.

58 Beckett and Keeling, "Rethinking Remediation," 217.

59 Author interview with Frank Winde, June 8, 2016.

60 Winde and de Villiers, "Uranium Contamination of Streams by Tailings Deposits," 811.

61 Coetzee, Wade, and Winde, "Reliance on Existing Wetlands," 62.

62 Coetzee, Wade, and Winde, "Reliance on Existing Wetlands," 62.

63 Winde and de Villiers, "Uranium Contamination of Streams by Tailings Deposits," 811.

64 Coetzee, Winde, and Wade, "An Assessment of Sources," v.

65 Winde and de Villiers, "The Nature and Extent of Uranium Contamination," 894.

66 Coetzee, Winde, and Wade, "An Assessment of Sources," 166.

67 See chapter 8 of Hecht, *Being Nuclear*, for a fuller discussion of the NNR.

68 Coetzee, Winde, and Wade, "An Assessment of Sources," xix.

69 Barthel, "Assessment of the Radiological Impact," 109.

70 Barthel, "Assessment of the Radiological Impact," 41.

71 Cram, "Becoming Jane." For a detailed discussion of the global and South African histories of radiation standards in mines, see chapter 6 of Hecht, *Being Nuclear*.

72 Barthel, "Assessment of the Radiological Impact," 45.

73 Du Toit, "Background Report on Communities at Risk," 21.

74 Adriana Stuijt, "SA Radioactive Stream—400,000 at High Risk," Rense .com, July 31, 2001, https://rense.com/general77/sarad.htm.

75   Joel Avni, "State Knew about Danger for 40 Years," *Sowetan Live*, July 24, 2007, https://www.sowetanlive.co.za/news/2007-07-24-state-knew-about -danger-for-40-years/.

76   Winde, "Uranium Pollution of Water," 13.

77   Steven Lang, "Radioactive Water, the Price of Gold," Inter Press Service, December 3, 2007, http://www.ipsnews.net/2007/12/environment-south -africa-radioactive-water-the-price-of-gold/.

78   Winde, "Uranium Pollution of Water," 14.

79   Environmental Management Section, "Radiation and Toxicity Concerns."

80   Du Toit, "Background Report on Communities at Risk," 36.

81   Du Toit, "Background Report on Communities at Risk," 11.

82   Du Toit, "Background Report on Communities at Risk," 57.

83   Du Toit, "Background Report on Communities at Risk," 64.

84   Du Toit, "Background Report on Communities at Risk," 69.

85   Liefferink, "Assessing the Past and the Present Role of the National Nuclear Regulator," 18.

86   Environmental Management Section, "Radiation and Toxicity Concerns."

87   The three studies they allowed were Barthel, "Assessment of the Radiological Impact"; Wade et al., "Tier 1 Risk Assessment"; Coetzee, Winde, and Wade, "An Assessment of Sources."

88   Winde, "Questions"; NNR, "Regulatory Strategy."

89   DWAF AND NNR, "Wonderfontein Catchment Area: Remediation Action Plan."

90   Winde, "Questions."

91   WHO, "Uranium in Drinking Water," 4.

92   Hecht, *Being Nuclear*.

93   Winde, "Uranium Pollution of Water," 3–5.

94   Winde, "Uranium Pollution of Water," 5.

95   For an overview of this literature, see Hecht, "Does Chernobyl Still Matter?"

96   Oreskes and Conway, *Merchants of Doubt*.

97   Hepler-Smith, "Molecular Bureaucracy."

98   Liefferink, "Classification of FSE's Activism Pertaining to AMD."

99   Liefferink, "Assessing the Past and the Present Role of the National Nuclear Regulator," 16.

100   National Environmental Management Act of 2010, available at https://www.dffe.gov.za/legislation/actsregulations.

101   Mtolo, "Effectiveness of Environmental Management Frameworks"; Maneli and Musundire, "Investigating the Impact"; Cilliers, "The Quality and Effectiveness of Environmental Management Frameworks."

102   Randfontein and Westonaria later merged into the Rand West City Local Municipality.

103   BKS, "West Rand District Municipality," chap. 2.

104   BKS, "West Rand District Municipality," 113.

105   Devex, "BKS Group," accessed February 1, 2022, https://www.devex.com /organizations/bks-group-46380.

106   BKS, "West Rand District Municipality," 113.

107   BKS, "West Rand District Municipality," 119–20.

108   BKS, "West Rand District Municipality," 112.

109   BKS, "West Rand District Municipality," 68–69.

110   Environomics, "Gauteng Provincial Environmental Management Framework."

111   McCarthy et al., "Mine Water Management," 2.

112   McCarthy et al., "Mine Water Management," iv.

113   McCarthy et al., "Mine Water Management," 13.

114   McCarthy et al., "Mine Water Management," 69.

115   McCarthy et al., "Mine Water Management," 78.

116   International Human Rights Clinic, "The Cost of Gold," 48.

117   Bobbins, "Acid Mine Drainage."

118   Winde, "Uranium Pollution of Water," 21.

THREE   THE INSIDE-OUT RAND

1   Katz, *The White Death*; McCulloch, "Mining Evidence."

2   Mpofu-Walsh, *The New Apartheid*, Kindle loc. 377.

3   Quoted in McCulloch, "Mining Evidence," 73.

4   Quoted in McCulloch, "Mining Evidence," 73.

5   Quoted in McCulloch, "Mining Evidence," 76, emphasis mine.

6   Livingston, *Debility and the Moral Imagination in Botswana*; Ledwaba and Sadiki, *Broke and Broken*.

7   All quotes from "1946 African Mineworkers Strike."

8   *Dying for Gold* was shot primarily in Lesotho, which over the decades had sent tens of thousands of men to the mines. It premiered in Bizana, in the heart of the Eastern Cape's Mpondoland, another major labor-sending area.

9   "Tshiamiso Trust—Compensation for Silicosis and TB," https://www .tshiamisotrust.com.

10   Mushai, "The Long Road to Compensation for Silicosis Sufferers."

11   "Tshiamiso Trust—Compensation for Silicosis and TB."

12   Molelekwa, "Silicosis Payouts Are 'Symbolic Justice.'"

13   Margaret Bourke-White, "South Africa," *Life*, September 18, 1950; Lichtenstein and Halpern, *Margaret Bourke-White and the Dawn of Apartheid*.

14   Quote at circa 38 minutes in Schadeberg, "Ernest Cole." For a fuller analysis of Cole's life and work, see Jansen, "Colored Black." For an account of the troubled archiving of Cole's photographs, see Smith, "The Mystery of Ernest Cole's Archive."

15   Bowker and Star, *Sorting Things Out*; Breckenridge, "The Book of Life"; Breckenridge, "Verwoerd's Bureau of Proof."

16   Cole, *House of Bondage*, 18–19.

17   Cole, *House of Bondage*, 23.

18   Gaule, "Ernest Cole."

19   Cole, *House of Bondage*, 23.

20   Dobler, "Waiting," 13.

21   Gaule, "Ernest Cole," 391. On the politics of waiting: Auyero and Ollion, "Faire patienter, c'est dominer"; Auyero, *Patients of the State*; Mujere, "Unemployment, Service Delivery and Practices of Waiting"; Stasik, Hänsch, and Mains, "Temporalities of Waiting in Africa."

22   Cole, *House of Bondage*, 24.

23   Abrahams, "Fancies Idle."

24   Lichtenstein, "Mining Nostalgia."

25   Gordimer, "The Witwatersrand," 20.

26   Gordimer, "The Witwatersrand," 21–25.

27   Gordimer, "The Witwatersrand," 25–26.

28   International Human Rights Clinic, "The Cost of Gold," 66.

29   Kootbodien et al., "Heavy Metal Contamination."

30   Quoted in Reichardt, "The Wasted Years," 51–52.

31   Reichardt, "The Wasted Years."

32   Reichardt, "The Wasted Years," 77–83.

33   Quoted in Reichardt, "The Wasted Years," 119.

34   Quoted in Reichardt, "The Wasted Years," 141.

35   Quoted in Reichardt, "The Wasted Years," 114.

36   Reichardt, "The Wasted Years," 580.

37   Reichardt, "The Wasted Years," 159.

38   Blight, "Destructive Mudflows."

39   Environmental Mining Processing Rehabilitation Consultancy, accessed August 8, 2022, https://www.emprconsultancy.com/.

40   Reichardt, "The Wasted Years," 584; Niekerk and Viljoen, "Causes and Consequences of the Merriespruit."

41   Niekerk and Viljoen, "Causes and Consequences of the Merriespruit," 201.

42   Niekerk and Viljoen, "Causes and Consequences of the Merriespruit," 201.

43   Maria Kielmas, "Mining Disaster Covered; South African Accident Calls Liability System into Question," *Business Insurance*, March 7, 1994, 51; "Judge Condemns Harmony 'Fiasco,'" *Mining Journal*, April 7, 1995, 253.

44   Butcher, "Making and Governing Unstable Territory." See also Bremner, "The Political Life of Rising Acid Mine Water."

45   Butcher, "Making and Governing Unstable Territory," 2194. She further argues that the relationships among activists, media, academics, artists, and think tanks played a key role in a "war of manoeuvre" that combined representational reframing with legal coercion.

46   Cane, *Civilising Grass*.

47   Le Roux, "Designing KwaThema."

48   Nieftagodien and Gaule, *Orlando West, Soweto*; Bonner, "The Politics of Black Squatter Movements"; Posel, "The Case for a Welfare State."

49   Verwoerd, "Policy of the Minister of Native Affairs," 43.

50   Bench Marks, "'Waiting to Inhale,'" 57.

51   Quoted in Beningfield, *The Frightened Land*, 204.

52   Beningfield, *The Frightened Land*, 204.

53   Bench Marks, "'Waiting to Inhale,'" v.

54   Hecht, *Being Nuclear*.

55   Jan Smuts, Union of South Africa, House of Assembly Debates, August 23, 1948, 734.

56   Daniel Malan, "Speech by the Honorable the Prime Minister on the Occasion of the Opening of the First Uranium Production Plant in the Union of South Africa," October 8, 1952, 2, INL 1/2/41 Ref 21/12, National Archives of South Africa.

57   West Rand Consolidated Mines Limited, "The Story of South Africa's First Uranium Production Plant," 1952, AB16/2546, National Archives (United Kingdom).

58   Reichardt, "The Wasted Years," 301.

59   Caryle Murphy, "It's Boom Time in South Africa as Gold Soars; Substantial Money for Expansion," *Washington Post*, February 17, 1980, F1.

60   The East Rand Gold and Uranium Company (ERGO) plant at Brakpan recovered gold, uranium, and pyrite from mine waste using a pyrite flotation process. In its first few years, ERGO processed around twenty megatons of sands and slimes per year. Ellis and Sweetwood, "Mineral Industry of the Republic of South Africa."

61   Murphy, "It's Boom Time in South Africa."

62   "ERGO to Be Reborn," *Mining Review Africa*, August 19, 2008, https://www.miningreview.com/top-stories/ergo-to-be-reborn/.

63   Butcher, "Making and Governing Unstable Territory," 2194.

64   "Australia's Mintails to Revive Mogale Gold," *Africa News*, May 25, 2006; John Beveridge, "Gold Rush Puts Investors in a Spin," *Herald Sun* (Australia), December 9, 2005. See also DRD Gold Limited, "ERGO: Operation Overview," accessed June 4, 2020, https://www.drdgold.com/our-business/ergo.

65   NewCo Gold, accessed June 4, 2020, https://www.newcogold.co.za/about-us/.

66   Eric Onstad, "There's (Still) Gold in Them Thar Mine Dumps!," *Business Day*, March 27, 2007, 17.

67   Olalde, "The Haunting Legacy."

68   Lumkile Mondi, personal communication, December 14, 2021.

69   O. S. Spadavecchia, "Pamodzi Launches $1,3bn Resources-Focused Private Equity Fund," *Mining Weekly*, August 3, 2007.

70   Charlotte Matthews, "Harmony Gold's Uranium Project Boost for West Rand's Fortunes," *Business Day*, December 20, 2007, 7.

71   Lumkile Mondi, personal communication, December 14, 2021. For more on the IDC, see Mondi and Roberts, "The Role of Development Finance."

72   Chipkin et al., *Shadow State*; Madonsela, "Critical Reflections on State Capture"; Martin and Solomon, "Understanding the Phenomenon of 'State Capture.'"

73   Reichardt, "The Wasted Years," 71.

74   Fourie and van der Walt, "Heritage Scoping Assessment," 26–27.

75   Quoted in Fourie and van der Walt, "Heritage Scoping Assessment," 28.

76   Reichardt, "The Wasted Years," 71.

77   Quoted in Fourie and van der Walt, "Heritage Scoping Assessment," 28.

78   Reichardt, "The Wasted Years," 103.

79   Gaule, "Stardust Artist's Statement," 11.

80   Gaule, "Stardust Artist's Statement," 2.

81   Gaule, "Stardust Artist's Statement," 4.

82   Gaule, "Stardust Artist's Statement," 30, 33.

83   Gevisser, "My Johannesburg."

84   Peterson, Gavua, and Rassool, *The Politics of Heritage in Africa*; Herwitz, *Heritage, Culture, and Politics in the Postcolony*; Witz, *Apartheid's Festival*; Witz, Rassool, and Minkley, "Repackaging the Past for South African Tourism"; Rassool, "The Rise of Heritage"; Rassool and Witz, "South Africa: A World in One Country."

85   Reid, "Dwelling in Displacement."

86   Madida, "Exploring the Conflicting Meanings."

87   Fourie and van der Walt, "Heritage Scoping Assessment," 35.

88   Hemann Niebuhr, quoted in van Heerden, "Death of a Mine Dump," 66.

89   Gaule, "Stardust Artist's Statement."

90   Bench Marks, "'Waiting to Inhale,'" 65.

91   Liefferink, "Assessing the Past and the Present Role of the National Nuclear Regulator," 19.

92   Van Heerden, "Death of a Mine Dump."

93   Mark Gevisser, "My Johannesburg," *Nation*, April 2, 2014.

94   Bobby Jordan, "Mines Plan Super-Dumps the Size of Neighbourhood," *Sunday Times*, August 23, 2009.

95   Sarah Hudleston, "Vaal Residents in Uproar over Authorization of Toxic Dump," *Business Day*, August 15, 2009.

96   Sheree Bega, "Farmers Unite against Proposed 'Super Dump,'" *Star*, March 27, 2010.

97   Sheree Bega, "Victims of Contaminated Water and Dust: 'Mine Waste Destroying Lives'; Farmers Fear for Their Lives and Livelihoods in Toxic Area," *Star*, October 30, 2010.

98   Sandstorm Gold Royalties, "Mine Waste Solutions," accessed August 8, 2022, https://www.sandstormgold.com/our-royalties/mws/.

1   Nkosinathi Sithole, SERI, meeting with author, July 1, 2016.

2   Residues of spatial racialization persist in language. Historically white ag-
     glomerations are typically called towns or cities, while historically Black
     agglomerations are still called townships—hence, Kagiso (township, popu-
     lation about 100,000) versus Krugersdorp (town, population about 86,000).

3   Mogale City Local Municipality, "The History of Mogale," accessed August
     8, 2022, https://www.mogalecity.gov.za/mogale-city/explore-our-city/.

4   Peterson et al., *The Politics of Heritage in Africa*; Witz, *Unsettled History*;
     Reid, "Dwelling in Displacement."

5   Khumalo, "Kagiso Historical Research Report." The study also aimed to
     identify historical markers for a future park and establish parameters for fu-
     ture heritage projects.

6   "The June 16 Soweto Youth Uprising."

7   Quoted in Khumalo, "Kagiso Historical Research Report," 18.

8   Figures taken from Frith, "Mogale City."

9   Phumla Patience Mjadu, "Founding Affidavit" in the Application of Ex Parte
     Mjadu: Federation for a Sustainable Environment v National Nuclear Regu-
     lator (South Gauteng High Court, Case No. 24611/2012), July 27, 2012, para.
     13–15.

10   Masuku, Langa, and Bruce, "The Community Work Programme in Kagiso."

11   All quotes from Masuku, Langa, and Bruce, "The Community Work Pro-
     gramme in Kagiso."

12   All quotes from Masuku, Langa, and Bruce, "The Community Work Pro-
     gramme in Kagiso."

13   Honwana, *Youth and Revolution in Tunisia*, coins the term *waithood* to de-
     scribe the lengthening purgatorial period between adolescence and adult-
     hood in much of Africa.

14   All quotes from Masuku, Langa, and Bruce, "The Community Work Pro-
     gramme in Kagiso."

15   The waste work of the Community Work Programmes represents only a
     small segment of a much wider range of formal and informal waste work.
     For a sophisticated entry into this literature, see Samson, "Accumulation by
     Dispossession and the Informal Economy."

16   Moloi, "When Gold Becomes a Nuisance"; Madihlaba, "The Environmen-
     tal Impact of Mining"; Glinski, "Environmental Justice in South African Law
     and Policy."

17   P. D. Toens, interview with author, Cape Town, South Africa, October 13,
     2003.

18   Toens, Stadler, and Wullschlege, "The Association of Groundwater Chemis-
     try and Geology with Atypical Lymphocytes."

19   Winde, "Uranium Pollution of Water," 12.

20   Phillips, "Surveillance Report of the Upper Wonderfonteinspruit," 2.

21 Phillips, "Surveillance Report of the Upper Wonderfonteinspruit," 26.

22 Sheree Bega, "Living in SA's Own Chernobyl," *Saturday Star*, January 8, 2011.

23 On knowing nature through labor, see White, *The Organic Machine*.

24 Bega, "Living in SA's Own Chernobyl."

25 Mogale City Local Municipality, EXCO meeting, January 19, 2011, 5–7. Report courtesy of M. Liefferink, personal collection.

26 C. O. Phillips (NNR) to Mr. D. Mashitisho (MCLM), 18 January 2011, re: Radiological Status of Tudor Shaft Informal Settlement, Annexure H to Mogale City Local Municipality, EXCO meeting, January 19, 2011.

27 Mogale City Local Municipality, EXCO meeting, January 19, 2011, 8.

28 Mogale City Local Municipality, EXCO meeting, January 19, 2011, 8.

29 Rasmussen, "Struggling for the City"; A. Johnson, "Post-apartheid Citizenship and the Politics of Evictions"; Wilhelm-Solomon, "Decoding Dispossession." More recent commentary: "Brutal Evictions Worsen South Africa's Urban Crisis" (editorial), *New Frame*, July 10, 2020.

30 Max Rambau, "Red Ants Relocate 35 Families," *SASDI Alliance* (blog), March 1, 2011, https://sasdialliance.org.za/2011/03/01/red-ants-relocate-35-families-to-mogale/.

31 Boyce Mkhize and Orion Phillips, "The NNR Report and Relocation of Residents in Tudor Shaft Informal Settlement," presentation to National Assembly Energy Committee, March 1, 2011, https://pmg.org.za/committee-meeting/12639/.

32 Phillips, "Surveillance Report of the Upper Wonderfonteinspruit."

33 Du Toit, "Background Report on Communities at Risk," 9–10.

34 Hecht, *Being Nuclear*.

35 Valentin, *The 2007 Recommendations of the International Commission on Radiological Protection*, 109.

36 Hecht, *Being Nuclear*.

37 Humby, "Environmental Justice and Human Rights."

38 Sheree Bega, "Residents Seek to Block Removal of Toxic Waste: The Area Is Home to 5,000 but Only 35 Families Have Been Moved by the Council," *Star* (South Africa), June 30, 2012.

39 Humby, "Environmental Justice and Human Rights," 104.

40 Mjadu, "Founding Affidavit."

41 NNR, Media Statement, "Objection to a 'One-Sided' Article Appearing in the *Saturday Star*," November 5, 2012.

42 Quoted in Kisting et al., "Case Study on Extractive Industries."

43 Mr. Jeffrey Ramoruti, meeting with Nkosinathi Sithole, SERI, July 1, 2016; undated handwritten memo from 2016. He asked for 8 million rands in royalties and damages for the use of his name. Courtesy of J. Ramoruti, personal collection.

44 Mjadu, "Founding Affidavit," 61.

45 "Residents Wait Years for Deeds," *Sowetan*, December 7, 2010.

46 International Human Rights Clinic, "The Cost of Gold," 71.

47 Solly Maphumulo, "Bloom Keen to Take Over as Gauteng Premier," *IOL*, July 29, 2013, https://www.iol.co.za/news/south-africa/gauteng/bloom-keen-to-take-over-as-gauteng-premier-1554162; Jack Bloom, "The ANC Don't Own This Place," *Politics Web*, February 6, 2012, https://www.politicsweb.co.za/politics/the-anc-dont-own-this-place.

48 Humby, "Environmental Justice and Human Rights"; Sherman, "Forgotten People."

49 Humby, "Environmental Justice and Human Rights," 107.

50 Dawid de Villiers, "Expert Opinion on How Best to Deal with the Tailings at Tudor Shaft Informal Settlement," November 16, 2015, Doc. No. SR-REP-011/2015, 2. Courtesy of SERI, private communication.

51 De Villiers, "Expert Opinion on How Best to Deal with the Tailings," 3.

52 Kromdraai Catchment, Minutes of the Wonderfontein-/Loop Spruit Meeting, June 23, 2015, Randfontein, 10. Courtesy of M. Liefferink, personal collection.

53 Sheree Bega, "An 'Environmental Catastrophe': NGO Seeks to Hold Mintails Group Accountable for Its Actions," *Star*, August 31, 2019.

54 Author visit to Tudor Shaft, July 4, 2016.

55 Author visit to Tudor Shaft, July 4, 2016.

56 Jeffrey Ramoruti, memo, personal papers, used with permission.

57 Sheree Bega, "Life in Toxic Wasteland No Child's Play," *Star*, March 14, 2016; Sheree Bega, "Trapped in a Nightmare of Hazardous Dust," *Star*, November 1, 2014.

58 Baloyi and Lubinga, "After Izimbizo, What Next?"

59 Jeff Radebe, "Minister Jeff Radebe: Mogale City Community Imbizo," South African Government, November 11, 2016, https://www.gov.za/speeches/community-imbizo-11-nov-2016-0000.

60 Angus Begg, "SA National Nuclear Regulator Neglects Radioactive Mine Dumps Threating the Health of Thousands," *Daily Maverick*, October 18, 2021.

61 Quoted in Barbara Maregele and Mary-Anne Gontsana, "Families Living Next to 'Toxic' Mine Dump Relocated," *Ground Up*, February 6, 2017.

62 "Mjadu and Others in re Federation for a Sustainable Environment (FSE) v National Nuclear Regulator and Others ('Tudor Shaft')," Socio-Economic Rights Institute of South Africa, accessed July 9, 2019, http://www.seri-sa.org/index.php/component/content/article?id=118:mjadu-and-others-in-re-federation-for-a-sustainable-environment-fse-v-national-nuclear-regulator-and-others-tudor-shaft.

63 Moshupya et al., "Assessment of Radon Concentration." Another study—also more rigorous than its predecessors—found high levels of arsenic and nickel in soil where informal settlement residents grew crops: Ngole-Jeme, Mpode, and Fantke, "Ecological and Human Health Risks Associated with Abandoned Gold Mine Tailings."

64   Moshupya et al., "Assessment of Radon Concentration."

65   Federation for a Sustainable Environment, "Integrated Annual Report 2018."

66   Larkin, *Tales from the City of Gold*; Jansson, *An Acid River Runs through It*; Tang and Watkins, "Ecologies of Gold."

67   The film essentially reproduced her toxic tours for an international audience; before the coronavirus pandemic shut everything down, it showed at festivals in Rotterdam, New York, Kosovo, and Warsaw.

68   Mark Olalde, "Mine Rehabilitation: Liquidation Allows Mintails to Shirk Environmental Liabilities," *Business Day*, August 21, 2018.

69   Mark Olalde and Andiswa Matikinca, "Directors Targeted for Mintails Mess," *Oxpeckers*, December 12, 2018, https://oxpeckers.org/2018/12/mintails-directors-targeted/.

70   Report of the Portfolio Committee on Mineral Resources on its oversight visit to North West and Gauteng on September 13–14, 2018, November 7, 2018.

71   Sheree Bega, "Mintails Mining Site Reduced to Rubble," *Saturday Star*, August 31, 2019.

72   "Zamas Hit West Rand Mine; Ransacking the Now-Liquidated Mintails' Mine Area May Have Dire Consequences," *Star*, June 1, 2019.

73   "Founding Affidavit in Federation for a Sustainable Environment v. Chief Directorate and Regional Manager for Gauteng for Mineral Resources and Others," Federation for a Sustainable Environment, May 29, 2019. Courtesy of M. Liefferink, personal collection.

74   Bega, "An 'Environmental Catastrophe.'"

75   Chris Barron, "Miner Embarks on Blueberry Farming," *Sunday Times*, June 13, 2021.

76   Lucky Masiga, "PAR Faces the Mintails Beast," *African Mining Online* (blog), October 1, 2021, https://www.africanmining.co.za/2021/10/01/par-faces-the-mintails-beast/.

77   Mariette Liefferink, "Report on the Site Visit on the 15th of December, 2021 of Amatshe Mining's Operations within Krugersdorp/Randfontein," Federation for a Sustainable Environment, January 3, 2022. Courtesy of M. Liefferink, personal collection.

FIVE LAND MINES

1   Sheree Bega, "Coal Mine Acidic Water Spillage into Wilge River System Kills 'All Life, Everything,'" *Mail and Guardian*, February 24, 2022, https://mg.co.za/environment/2022-02-24-coal-mine-acidic-water-spillage-into-wilge-river-system-kills-all-life-everything/.

2   Sheree Bega, "Eskom Is the World's Worst Polluting Power Company by SO_2," *Mail and Guardian*, October 6, 2021.

3   Mbembe and Shread, "The Universal Right to Breathe."

4   GDARD, "Study on Reclamation and Rehabilitation."

5   Altman, "Time-Bombing the Future," drawing on geneticist Hermann Muller's 1948 description of nuclear weapons tests.

6   Levin et al., "Overcoming the Tragedy of Super Wicked Problems."

7   Butcher, "Making and Governing Unstable Territory."

8   Butcher, "Making and Governing Unstable Territory."

9   A. Alexander, "Democracy Dispossessed." For more on this theme, see von Schnitzler, *Democracy's Infrastructure*; Park and Donovan, "Between the Nation and the State."

10   Manungufala, Sabiti-Kalule, and Aucamp, "Investigation into the Slimes Dams, Mine Dumps and Landfills." The authors worked for Tswelopele Environmental, one of many consulting companies that assessed environmental conditions around proposed sites of development.

11   GDARD, "Gauteng Mine Residue Area Strategy," 2012.

12   Harber and Joseph, "Institutionalising the Gauteng City-Region."

13   Jacobs, *Cities and the Wealth of Nations*.

14   Scott, *Global City-Regions*, 4.

15   Harber and Joseph, "Institutionalising the Gauteng City-Region."

16   All quotes from Gauteng City-Region Observatory, "Vision, Mandate and Core Roles," accessed August 10, 2021, https://www.gcro.ac.za/about /about-the-gcro/.

17   Gauteng City-Region Observatory, "GOVERNING THE GCR," accessed January 24, 2023, https://www.gcro.ac.za/research/research-themes/detail /governing-gcr/.

18   Ballard et al., "Scale, Belonging and Exclusion in Gauteng." Publications elaborating on these themes include Ballard et al., "Scale of Belonging"; Ballard, Jones, and Ngwenya, "Trickle-Out Urbanism"; Parker, Hamann, and de Kadt, "Accessing Quality Education in Gauteng"; Mkhize, "Rescaling Municipal Governance in Gauteng."

19   R. Moore, "Taking the Gap."

20   McCarthy, "The Decanting of Acid Mine Water."

21   McCarthy, "The Decanting of Acid Mine Water," viii.

22   Bobbins, "Acid Mine Drainage," 9.

23   Bobbins and Trangoš, *Mining Landscapes of the Gauteng City-Region*, 161.

24   Phasha, *A City on a Hill*, 7.

25   Potšišo Phasha, personal communication, November 13, 2021.

26   Phasha, personal communication, December 14, 2021.

27   Chakwizira, "Low-Income Housing Backlogs"; Marutlulle, "A Critical Analysis of Housing Inadequacy."

28   Quoted in Ballard and Rubin, "A 'Marshall Plan' for Human Settlements," 14.

29   Natalie Greve, "Megaproject Strategy to Permanently Alter Housing Development Landscape—Makhura," *Engineering News*, July 4, 2015, https://www.engineeringnews.co.za/article/megaproject-strategy-to-permanently-alter-housing-development-landscape-makhura-2015-04-07/rep_id:4136.

30   Wright, *The Politics of Design*.

31   Ballard and Rubin, "A 'Marshall Plan' for Human Settlements."

32   Quote in Greve, "Megaproject Strategy to Permanently Alter."

33   Calgro M3 Group, "Fleurhof," accessed August 23, 2021, https://calgrom3.com/index.php/fleurhof.

34   Oliver Balch, "Radioactive City: How Johannesburg's Townships Are Paying for Its Mining Past," *Guardian*, July 6, 2015.

35   John Moodley, "Gauteng's Housing Plan Echoes Sentiments of the Past," *Sowetan*, July 11, 2017.

36   Charlton, "Poverty, Subsidised Housing and Lufhereng."

37   Le Roux, "Designing KwaThema."

38   Charlton, "Poverty, Subsidised Housing and Lufhereng."

39   Nuttall and Mbembe, *Johannesburg*; Bremner, *Writing the City into Being*; Harrison et al., *Changing Space, Changing City*; Falkof and Van Staden, *Anxious Joburg*; Nieftagodien and Gaule, *Orlando West, Soweto*; Vladislavić, *Portrait with Keys*; Vladislavić, *The Restless Supermarket*; Beningfield, *The Frightened Land*; Murray, *Panic City*; Murray, *City of Extremes*.

40   Wambecq et al., "From Mine to Life Cycle Closure"; Toffah, "Mines of Gold, Mounds of Dust"; Lyssens and Juwet, "Sculpting the Hill." Special thanks to Hannah le Roux for her guidance and insights into this material.

41   City of Johannesburg, "Nodal Review Policy 2019/20."

42   City of Johannesburg et al., "Spatial Development Framework 2040," 103.

43   This had been known as the Informal Backyard Enablement Programme. City of Johannesburg et al., "Spatial Development Framework 2040," 140.

44   City of Johannesburg et al., "Spatial Development Framework 2040," 114.

45   Plan Associates, "About Us," accessed February 9, 2022, https://planassociates.co.za/about-us/.

46   This despite citing Markus Reichardt's thesis on the chamber's Vegetation Unit (see chapter 2).

47   Pretorius, "West Mining Belt."

48   Special thanks to Neil Overy for his assistance with the research and analysis in this section.

49   Alberts et al., "Complexities with Extractive Industries Regulation."

50   Humby, "Environmental Justice and Human Rights."

51   Humby, "Facilitating Dereliction?"

52   Watson and Olalde, "The State of Mine Closure."

53   Humby, "Environmental Justice and Human Rights," 461.

54 Provision 30(a) of Minerals and Petroleum Resources Development Amendment Bill in GG 36523 of May 31, 2013, Republic of South Africa, 20.

55 Humby, "The Spectre of Perpetuity," 463.

56 Department of Environmental Affairs, "Government's One Environmental System Commences," December 9, 2014, https://www.dffe.gov.za/media release/oneenvironmentalsystem.

57 Humby, "'One Environmental System.'"

58 Centre for Environmental Rights, "Time Is of the Essence: Proposed New Rules Slash Timeframes for EIA," October 2, 2014, https://cer.org.za/news /time-is-of-the-essence-proposed-new-rules-slash-timeframes-for-eia.

59 Alberts et al., "Complexities with Extractive Industries Regulation."

60 South African Human Rights Commission, "National Hearing on the Underlying Socio-economic Challenges," 74.

61 Shivamba, "Mining for Sustainable Development."

62 Milaras, Ahmed, and McKay, "Mine Closure in South Africa."

63 Power, *The Audit Society*; Strathern, *Audit Cultures*; Haufler, "Disclosure as Governance." On baselines, see Barandiarán, "Documenting Rubble to Shift Baselines."

64 Milaras, Ahmed, and McKay, "Mine Closure in South Africa."

65 For a recent example: Mine Closure, Fourteenth International Conference, accessed February 2, 2022, https://mineclosure2021.com/virtual -mineclosure/.

66 Steffen, Robertson and Kirsten (SRK), for example, began in 1974 as a mining consultancy. In the 1990s it expanded its environmental services; today it has a global presence with forty-five offices on six continents. "Our specialists are leaders in fields such as due diligence, feasibility studies, mine waste and water management, permitting, and mine closure," SRK's corporate profile proudly proclaims. "The industry background of many of our staff ensures that our advice is not only technically sound but also thoroughly practical." SRK Consulting, "Corporate Profile: An Overview of Who We Are," accessed January 24, 2023, https://www.kz.srk.com/sites/default /files/File/Brochure/srk-corporate-profile-2017-a4.pdf.

67 Shivamba, "Mining for Sustainable Development."

68 One exception is Kongiwe Environmental, founded in 2016, which posts its mammoth reports online and offers readers the chance to register as "Interested and Affected Parties" directly. Most firms do not do this. Kongiwe Environmental, "Public Documents," accessed August 25, 2021, http://www .kongiwe.co.za/publications-view/public-documents/.

69 Watson and Olalde, "The State of Mine Closure," 642.

70 This via legislation such as the Traditional Leadership and Governance Framework Act (2003) and the more recent Traditional and Khoi-San Leadership Act (2019).

71 Mnwana, "A Radical Mineral Policy with Retribalisation?"; Mnwana, "Democracy, Development and Chieftaincy"; Mnwana, "'Custom' and Fractured

'Community'"; Mnwana and Bowman, "Land, Conflict and Radical Distributive Claims"; Buthelezi, Skosana, and Vale, *Traditional Leaders in a Democracy*.

72 Bowman, "Black Economic Empowerment Policy."

73 Mpofu-Walsh, *The New Apartheid*, Kindle loc. 1516.

74 Mpofu-Walsh, *The New Apartheid*.

75 Quoted in Mnwana and Bowman, "Land, Conflict and Radical Distributive Claims," 8.

76 Quoted in Oesi, *Black Lives Matter*; Lloyd Gedye, "'Black Lives Matter' Explores Marikana Residents' Fight against a Mining Company, a Chief and a Municipality," *Mail and Guardian*, July 4, 2016.

77 Special thanks to Tara Weinberg for her research on this section.

78 DMR, "Presentation to the Portfolio Committee of Mineral Resources," February 22, 2017. Courtesy of M. Liefferink, personal collection.

79 For example, by making the most of a report by Bonnie Docherty of Harvard Law School's International Human Rights Clinic on the consequences of gold mining on the Rand. International Human Rights Clinic, "The Cost of Gold."

80 Bega, "Life in Toxic Wasteland No Child's Play," *The Star*, March 14, 2015.

81 South African Human Rights Commission, "National Hearing on the Underlying Socio-economic Challenges."

82 T. G. Moonsamy to Mariette Liefferink, "Response to Request for Access to Record of Public Body in Terms of the Promotion of Access to Information Act, 2000 (Act No 2 of 2000)," September 28, 2021. Courtesy of M. Liefferink, personal collection.

83 Mariette Liefferink, personal communication, January 29, 2022.

84 Mariette Liefferink to Jonas Sibanyoni, "Status of the Implementation and Enforcement of the SAHRC's Directives in Terms of the SAHRC Mining Communities Report (2016)," January 16, 2022. Courtesy of M. Liefferink, personal collection.

85 CSIR, "A CSIR Perspective on South Africa's Post-mining Landscape," emphasis mine.

86 Livingston, *Self-Devouring Growth*.

CONCLUSION

1 Mpofu-Walsh, *The New Apartheid*, Kindle loc. 2862.

2 Mudd, "Global Trends in Gold Mining"; Mudd, "The Limits to Growth and 'Finite' Mineral Resources"; Cooper et al., "Humans Are the Most Significant."

3 Hufty, "Abandoned Mines."

4 United Nations Environment Programme, "Abandoned Mines," 15.

5   They've been making this claim with particular fervor since the appearance of the Club of Rome's *Limits to Growth* (*LTG*) in 1972. For recent analysis of the ongoing tensions over the *LTG* thesis, see Nørgård et al., "The History of *The Limits to Growth*"; "Are There Limits to Economic Growth?" For a take on mining and *LTG*, see Mudd, "The Limits to Growth and 'Finite' Mineral Resources." For a more recent, detailed critique of the worldwide fascination with growth from someone who also objected to the modeling in *LTG*, see Smil, *Growth*.

6   Mills, *The Racial Contract.*

7   Critic Mark Fisher sees this as a form of "capitalist realism," one that depicts any "hope that these forms of suffering could be eliminated . . . as naive utopianism." Fisher, *Capitalist Realism*, 16.

8   Asif Siddiqi, personal communication, February 8, 2022. See also Bhimull et al., "Systemic and Epistemic Racism."

9   The relationships among formal and informal economic, labor, and housing practices have been widely studied. For a quick overview, see Hart, "On the Informal Economy." For an analysis of how matters of in/formality play out in geology and mining, see D'Avignon, *A Ritual Geology*. For an analysis of how these matters play out among waste reclaimers in South Africa, see Samson, "Accumulation by Dispossession."

10   Le Roux and Hecht, "Bad Earth."

11   Hecht, *Being Nuclear.*

12   On austerity, see Tousignant, *Edges of Exposure*; Park and Donovan, "Between the Nation and the State"; Grace, *African Motors*; Grace, "Poop."

13   In the US, for example, members of Congress can buy and sell individual stocks, leading some to submit to the temptations of insider trading. Not to mention the disastrous Supreme Court ruling in the *Citizens United* case, which essentially legalized corporate influence on politicians.

14   Special thanks to Ellen Poteet, whose remarks helped me think through this part of the argument. See also Mamdani, *Neither Settler nor Native*.

15   Card, "Geographies of Racial Capitalism."

BIBLIOGRAPHY

Abrahams, Peter. "Fancies Idle." In *You Better Believe It: Black Verse in English from Africa, the West Indies, and the United States*, edited by Paul Breman, 170–71. Baltimore, MD: Penguin Books, 1973.

Adas, Michael. *Machines as the Measure of Men: Science, Technology, and Ideologies of Western Dominance*. Ithaca, NY: Cornell University Press, 1989.

Adler, Rebecca A., Marius Claassen, Linda Godfrey, and Anthony R. Turton. "Water, Mining and Waste: An Historical and Economic Perspective on Conflict Management in South Africa." *Economics of Peace and Security Journal* 2, no. 2 (2007): 32–41.

Agricola, Georgius. *De re metallica*. Translated by Herbert Clark Hoover and Lou Henry Hoover. New York: Dover, 1950. https://www.gutenberg.org/files/38015/38015-h/38015-h.htm.

Alberts, R., J. A. Wessels, A. Morrison-Saunders, M. P. McHenry, A. Rita Sequeira, H. Mtegha, and D. Doepel. "Complexities with Extractive Industries Regulation on the African Continent: What Has 'Best Practice' Legislation Delivered in South Africa?" *Extractive Industries and Society* 4, no. 2 (April 1, 2017): 267–77.

Alexander, Amanda. "Democracy Dispossessed: Land, Law and the Politics of Redistribution in South Africa." PhD diss., Columbia University, 2016.

Alexander, Neville. *One Azania, One Nation: The National Question in South Africa*. Edited by No Sizwe. London: Zed, 1979.

Alexander, Neville. *An Ordinary Country: Issues in Transition from Apartheid to Democracy in South Africa*. New York: Berghahn, 2003.

Altman, Rebecca G. "Time-Bombing the Future." *Aeon*, January 2, 2019.

"Are There Limits to Economic Growth? It's Time to Call Time on a 50-Year Argument." Editorial. *Nature* 603, no. 7901 (March 16, 2022).

Ariza, Luis Miguel. "River of Vitriol." *Scientific American* 279, no. 3 (1998): 26–28.

Asafu-Adjaye, John, Linus Blomqvist, Stewart Brand, Barry Brook, Ruth DeFries, Erle Ellis, Christopher Foreman, et al. "An Ecomodernist Manifesto." 2015. http://www.ecomodernism.org.

Auf der Heyde, T. "South African Nuclear Policy since 1993: Too Hot to Handle?" Paper presented at National Nuclear Technology Conference, September 6–9, 1998, South Africa. Chamber of Mines archives.

Auyero, Javier. *Patients of the State: The Politics of Waiting in Argentina*. Durham, NC: Duke University Press, 2012.

Auyero, Javier, and Étienne Ollion. "Faire patienter, c'est dominer: Le pouvoir, l'État et l'attente." *Actes de la recherche en sciences sociales* 226–27, no. 1 (May 3, 2019): 120–25.

Ballard, Richard, Gareth A. Jones, and Makale Ngwenya. "Trickle-Out Urbanism: Are Johannesburg's Gated Estates Good for Their Poor Neighbours?" *Urban Forum* 32, no. 2 (June 2021): 165–82.

Ballard, Richard, Ngaka Mosiane, Alexandra Parker, Christian Hamann, Julia de Kadt, Siân Butcher, and Sandiswa Mapukata. "Scale, Belonging and Exclusion in Gauteng." Johannesburg: GCRO, n.d.

Ballard, Richard, Alexandra Parker, Siân Butcher, Julia de Kadt, Christian Hamann, Kate Joseph, Sandiswa Mapukata, Thembani Mkhize, Ngaka Mosiane, and Lukas Spiropoulos. "Scale of Belonging: Gauteng 30 Years after the Repeal of the Group Areas Act." *Urban Forum* 32, no. 2 (June 2021): 131–39. https://doi.org/10.1007/s12132-021-09429-5.

Ballard, Richard, and Margot Rubin. "A 'Marshall Plan' for Human Settlements: How Megaprojects Became South Africa's Housing Policy." *Transformation: Critical Perspectives on Southern Africa* 95, no. 1 (2017): 1–31.

Ballim, Faeeza. *Apartheid's Leviathan: Electricity and the Power of Technological Ambivalence*. Athens: Ohio University Press, 2023.

Baloyi, Mahlatse L., and Elizabeth N. Lubinga. "After Izimbizo, What Next? A Participatory Development Communication Approach to Analysing Feedback by the Limpopo Provincial Government to Its Citizens." *Journal for Transdisciplinary Research in Southern Africa* 13, no. 1 (April 26, 2017): 7.

Barandiarán, Javiera. "Documenting Rubble to Shift Baselines: Environmental Assessments and Damaged Glaciers in Chile." *Environment and Planning E: Nature and Space* 3, no. 1 (2020): 58–75.

Barthel, R. "Assessment of the Radiological Impact of the Mine Water Discharges to Members of the Public Living around Wonderfonteinspruit Catchment Area." BSA Project No. 0607-03, BS Associates, Bedfordview, South Africa, 2007.

Bauman, Zygmunt. *Wasted Lives: Modernity and Its Outcasts*. Cambridge: Polity, 2004.

Beckett, Caitlynn, and Arn Keeling. "Rethinking Remediation: Mine Reclamation, Environmental Justice, and Relations of Care." *Local Environment* 24, no. 3 (March 4, 2019): 216–30.

Beinart, William, and Saul Dubow. *The Scientific Imagination in South Africa: 1700 to the Present*. Cambridge: Cambridge University Press, 2021.

Bench Marks. "'Waiting to Inhale': A Survey of Household Health in Four Mine-Affected Communities." Policy Gap 12. Bench Marks Foundation, 2018. https://www.bench-marks.org.za/policy-gap-12/.

Beningfield, Jennifer. *The Frightened Land: Landscape and Politics in South Africa in the Twentieth Century*. New York: Routledge, 2006.

Bhimull, Chandra, Gabrielle Hecht, Edward Jones-Imhotep, C. Chakanetsa Mavhunga, Lisa Nakamura, and Asif Siddiqi. "Systemic and Epistemic Racism in the History of Technology." *Technology and Culture* 63, no. 4 (October 2022): 935–52.

BKS (Pty). "West Rand District Municipality: Environmental Management Framework, Revision 2." June 2013.

Blight, G. E. "Destructive Mudflows as a Consequence of Tailings Dyke Failures." *Proceedings of the Institution of Civil Engineering—Geotechnical Engineering* 125, no. 1 (1997): 9–18.

Bobbins, Kerry. "Acid Mine Drainage and Its Governance in the Gauteng City-Region." GCRO Occasional Paper 10. Johannesburg: GCRO, 2015. https://wiredspace.wits.ac.za/handle/10539/18835.

Bobbins, Kerry, and Guy Trangoš. *Mining Landscapes of the Gauteng City-Region*. Johannesburg: GCRO, 2018.

Bonner, Philip. "The Politics of Black Squatter Movements on the Rand, 1944–1952." *Radical History Review*, nos. 46–47 (May 1990): 89–115.

Boudia, Soraya, Angela N. H. Creager, Scott Frickel, Emmanuel Henry, Nathalie Jas, Carsten Reinhardt, and Jody A. Roberts. "Residues: Rethinking Chemical Environments." *Engaging Science, Technology, and Society* 4 (June 28, 2018): 165–78.

Boudia, Soraya, Angela N. H. Creager, Scott Frickel, Emmanuel Henry, Nathalie Jas, Carsten Reinhardt, and Jody A. Roberts. *Residues: Thinking through Chemical Environments*. New Brunswick, NJ: Rutgers University Press, 2022.

Bowker, Geoffrey C., and Susan Leigh Star. *Sorting Things Out: Classification and Its Consequences*. Cambridge, MA: MIT Press, 1999.

Bowman, Andrew. "Black Economic Empowerment Policy and State-Business Relations in South Africa: The Case of Mining." *Review of African Political Economy* 46, no. 160 (April 3, 2019): 223–45.

Braun, Lundy. *Breathing Race into the Machine: The Surprising Career of the Spirometer from Plantation to Genetics*. Minneapolis: University of Minnesota Press, 2014.

Breckenridge, Keith. "The Book of Life: The South African Population Register and the Invention of Racial Descent, 1950–1980." *Kronos* 40, no. 1 (November 1, 2014): 225–40.

Breckenridge, Keith. "Made in America: Progressive Mining Engineers and the Origins of Corporate Capitalism in South Africa." Unpublished manuscript, 2007. Courtesy of the author.

Breckenridge, Keith. "'Money with Dignity': Migrants, Minelords and the Cultural Politics of the South African Gold Standard Crisis, 1920–33." *Journal of African History* 36, no. 2 (January 1, 1995): 271–304.

Breckenridge, Keith. "Verwoerd's Bureau of Proof: Total Information in the Making of Apartheid." *History Workshop Journal* 59, no. 1 (February 10, 2006): 83–108.

Bremner, Lindsay. "The Political Life of Rising Acid Mine Water." *Urban Forum* 24, no. 4 (December 2013): 463–83.

Bremner, Lindsay. *Writing the City into Being: Essays on Johannesburg, 1998–2008*. Edited by Bronwyn Law-Viljoen. Johannesburg: Fourthwall, 2010.

Bullard, Robert D. *Unequal Protection: Environmental Justice and Communities of Color*. Oakland, CA: Sierra Club Books, 1994.

Bullard, Robert D., and Beverly Wright. *The Wrong Complexion for Protection: How the Government Response to Disaster Endangers African American Communities*. New York: New York University Press, 2012.

Butcher, Siân. "Making and Governing Unstable Territory: Corporate, State and Public Encounters in Johannesburg's Mining Land, 1909–2013." *Journal of Development Studies* 54, no. 12 (December 2, 2018): 2186–209.

Buthelezi, Mbongiseni, Dineo Skosana, and Beth Vale, eds. *Traditional Leaders in a Democracy: Resources, Respect and Resistance*. Johannesburg: Mapungubwe Institute for Strategic Reflection, 2019.

Cane, Jonathan. *Civilising Grass: The Art of the Lawn on the South African Highveld*. Johannesburg: Wits University Press, 2019.

Card, K., dir. "Geographies of Racial Capitalism with Ruth Wilson Gilmore." Antipode Foundation, 2020. https://antipodeonline.org/geographies-of-racial-capitalism/.

Chakwizira, James. "Low-Income Housing Backlogs and Deficits 'Blues' in South Africa. What Solutions Can a Lean Construction Approach Proffer?" *Journal of Settlements and Spatial Planning* 10, no. 2 (December 31, 2019): 71–88.

Charlton, Sarah. "Poverty, Subsidised Housing and Lufhereng as a Prototype Megaproject in Gauteng." *Transformation: Critical Perspectives on Southern Africa* 95, no. 1 (2017): 85–110.

Chipkin, Ivor, Mark Swilling, Haroon Bhorat, Mzukisi Qobo, Sikhulekile Duma, Lumkile Mondi, Camaren Peter, Mbongiseni Buthelezi, Hannah Friedenstein, and Nicky Prins. *Shadow State: The Politics of State Capture*. Johannesburg: Wits University Press, 2018.

Chirikure, Shadreck. *Metals in Past Societies*. New York: Springer International, 2015. https://doi.org/10.1007/978-3-319-11641-9.

Chirikure, Shadreck. "Metals in Society: Iron Production and Its Position in Iron Age Communities of Southern Africa." *Journal of Social Archaeology* 7, no. 1 (February 2007): 72–100.

Cilliers, D. P. "The Quality and Effectiveness of Environmental Management Frameworks (EMF) in South Africa." PhD diss., North-West University, 2015.

City of Johannesburg. "Nodal Review Policy 2019/20: City Transformation and Spatial Planning." February 2020.

City of Johannesburg Metropolitan Municipality in collaboration with Iyer Urban Design, UN Habitat, Urban Morphology and Complex Systems Institute and the French Development Agency. "Spatial Development Framework 2040." City of Johannesburg: Department of Development Planning, 2016.

Coase, R. H. "The Problem of Social Cost." *Journal of Law and Economics* 3 (October 1960): 1–44.

Coetzee, Henk, et al. "Uranium and Heavy Metals in Sediments in a Dam on the Farm Blaauwbank." Pretoria: Council for Geoscience, 2002.

Coetzee, Henk, Peter Wade, and Frank Winde. "Reliance on Existing Wetlands for Pollution Control around the Witwatersrand Gold/Uranium Mines of South Africa—Are They Sufficient?" In *Uranium in the Aquatic Environment*, edited by Broder J. Merkel, Britta Planer-Friedrich, and Christian Wolkersdorfer, 59–64. Berlin: Springer, 2002.

Coetzee, Henk, F. Winde, and P. W. Wade. "An Assessment of Sources, Pathways, Mechanisms and Risks of Current and Potential Future Pollution of Water and Sediments in Gold-Mining Areas of the Wonderfonteinspruit Catchment: Report to the Water Research Commission." Water Research Commission, Report 1214, January 2006.

Cole, Ernest. *House of Bondage*. New York: Random House, 1967.

Cooper, Anthony H., Teresa J. Brown, Simon J. Price, Jonathan R. Ford, and Colin N. Waters. "Humans Are the Most Significant Global Geomorphological Driving Force of the 21st Century." *Anthropocene Review* 5, no. 3 (December 1, 2018): 222–29.

Cram, Shannon. "Becoming Jane: The Making and Unmaking of Hanford's Nuclear Body." *Environment and Planning D: Society and Space* 33, no. 5 (October 1, 2015): 796–812.

CSIR. "A CSIR Perspective on South Africa's Post-mining Landscape." Pretoria: Council for Scientific and Industrial Research, 2019.

D'Avignon, Robyn. "Minerals." *Somatosphere*, December 11, 2017.

D'Avignon, Robyn. "Primitive Techniques: From 'Customary' to 'Artisanal' Mining in French West Africa." *Journal of African History* 59, no. 2 (2018): 179–97.

D'Avignon, Robyn. *A Ritual Geology: Gold and Subterranean Knowledge in Savanna West Africa*. Durham, NC: Duke University Press, 2022.

D'Avignon, Robyn. "Spirited Geobodies: Producing Subterranean Property in Nineteenth-Century Bambuk, West Africa." *Technology and Culture* 61, no. 2 (April 2020): S20–S48.

Death, Carl. "Troubles at the Top: South African Protests and the 2002 Johannesburg Summit." *African Affairs* 109, no. 437 (October 1, 2010): 555–74.

Dillon, Lindsey. "Race, Waste, and Space: Brownfield Redevelopment and Environmental Justice at the Hunters Point Shipyard." *Antipode* 46, no. 5 (November 2014): 1205–21.

Dlamini, Jacob S. T. *Safari Nation: A Social History of Kruger National Park*. Athens: Ohio University Press, 2020.

Dobler, Gregor. "Waiting: Elements of a Conceptual Framework." *Critical African Studies* 12, no. 1 (January 2, 2020): 10–21.

Dubow, Saul. *A Commonwealth of Knowledge: Science, Sensibility, and White South Africa, 1820–2000*. Oxford: Oxford University Press, 2006.

Dubow, Saul. *Scientific Racism in Modern South Africa*. Cambridge: Cambridge University Press, 1995.

Dugard, Jackie, and Anna Alcaro. "A Rights-Based Examination of Residents' Engagement with Acute Environmental Harm across Four Sites on South Africa's Witwatersrand Basin." *Social Research* 79, no. 4 (winter 2012): 931–56.

du Toit, J. S. "Background Report on Communities at Risk within Mogale City Local Municipality Affected by Mining Related Activities, with Special Reference to Radiation and Toxicity." Krugersdorp: Mogale City Local Municipality, September 2007.

DWAF (Department of Water Affairs and Forestry) and NNR (National Nuclear Regulator). "Wonderfontein Catchment Area: Remediation Action Plan. Radioactive Contamination Specialist Task Team Report on Site Visits and Recommended Actions." Rev. 1.0, October 24, 2008.

Ebron, Paulla A. "Slave Ships Were Incubators for Infectious Diseases." In *Feral Atlas: The More-Than-Human Anthropocene*, edited by Anna L. Tsing, Jennifer Deger, Alder Keleman Saxena, and Feifei Zhou. Stanford, CA: Stanford University Press, 2020. https://feralatlas.supdigital.org/poster/slave-ships-were-incubators-for-infectious-diseases.

Edwards, Paul N. "Entangled Histories: Climate Science and Nuclear Weapons Research." *Bulletin of the Atomic Scientists* 68, no. 4 (July 1, 2012): 28–40.

Edwards, Paul N. "The Mechanics of Invisibility: On Habit and Routine as Elements of Infrastructure." In *Infrastructure Space*, edited by Ilka Ruby and Andreas Ruby, 327–36. Berlin: Ruby, 2017.

Edwards, Paul N., and Gabrielle Hecht. "History and the Technopolitics of Identity: The Case of Apartheid South Africa." *Journal of Southern African Studies* 36, no. 3 (September 2010): 619–39.

Ellis, Miller W., and Charles W. Sweetwood. "The Mineral Industry of the Republic of South Africa." *Minerals Yearbook Area Reports: International* 3 (1977): 825–48.

Environmental Management Section. "Radiation and Toxicity Concerns: Informal Settlements in the Tudor Shaft Area." Mogale City Local Municipality: Integrated Environmental Management, January 19, 2011.

Environomics. "Gauteng Provincial Environmental Management Framework (GPEMF): Environmental Management Framework Report." Gauteng Province: Gauteng Department of Agriculture and Rural Development, 2014.

Falkof, Nicky, and Cobus Van Staden, eds. *Anxious Joburg: The Inner Lives of a Global South City*. Johannesburg: Wits University Press, 2020.

Fanon, Frantz. *The Wretched of the Earth*. New York: Grove, 1968.

Federation for a Sustainable Environment. "Integrated Annual Report 2018." Rivonia, 2018.

Ferdinand, Malcom. *Une écologie décoloniale: Penser l'écologie depuis le monde caribéen*. Paris: Le Seuil, 2019.

Finney, Carolyn. *Black Faces, White Spaces: Reimagining the Relationship of African Americans to the Great Outdoors*. Chapel Hill: University of North Carolina Press, 2014.

Fisher, Mark. *Capitalist Realism: Is There No Alternative?* Winchester, UK: Zero, 2009.

Fortun, Kim. *Advocacy after Bhopal: Environmentalism, Disaster, New Global Orders*. Chicago: University of Chicago Press, 2001.

Fourie, Wouter, and Jaco van der Walt. "Heritage Scoping Assessment for the Top Star Dump Mining Project—Crown Gold Recoveries." Matakoma Heritage Consultants, February 10, 2006. Unpublished report.

Fressoz, Jean-Baptiste. *L'apocalypse joyeuse: Une histoire du risque technologique*. L'Univers Historique. Paris: Seuil, 2012.

Friede, H. M. "Iron Age Mining in the Transvaal." *Journal of the South African Institute of Mining and Metallurgy* (April 1980): 156–65.

Frith, Adrian. "Mogale City: Local Municipality 763 from Census 2011." Census 2011. Accessed March 23, 2022. https://census2011.adrianfrith.com/place/763.

Frost, M. "Report on an Investigation into Mine Effluents." Johannesburg: Chamber of Mines Research Organisation, 1957.

Funke, Nikki, Shanna Nienaber, and Christine Gioia. "An Interest Group at Work: Environmental Activism and the Case of Acid Mine Drainage on Johannesburg's West Rand." Africa Institute of South Africa, 2012.

Gaule, Sally. "Ernest Cole: The Perspective of Time." *Safundi* 18, no. 4 (October 2, 2017): 380–99.

Gaule, Sally. "Stardust Artist's Statement." November 8—December 16, 2012, Goethe on Main Gallery. Courtesy of the artist.

GDARD. "Gauteng Mine Residue Area Strategy." Johannesburg: Gauteng Department of Agriculture and Rural Development, 2012.

GDARD. "Study on Reclamation and Rehabilitation of Mine Residue Areas for Development Purposes: Phase II (Strategy and Implementation Plan)." Johannesburg: Gauteng Department of Agriculture and Rural Development, 2012.

Gevisser, Mark. *Lost and Found in Johannesburg: A Memoir*. New York: Farrar, Straus and Giroux, 2014.

Gidwani, Vinay, and Rajyashree N. Reddy. "The Afterlives of 'Waste': Notes from India for a Minor History of Capitalist Surplus." *Antipode* 43, no. 5 (2011): 1625–58.

Gilmore, Ruth Wilson. *Golden Gulag: Prisons, Surplus, Crisis, and Opposition in Globalizing California*. Berkeley: University of California Press, 2007.

Glinski, Carola. "Environmental Justice in South African Law and Policy." *Verfassung in Recht und Übersee* 36, no. 1 (2003): 49–74.

Global Justice Now. "Honest Accounts 2017: How the World Profits from Africa's Wealth." May 24, 2017. https://www.globaljustice.org.uk/resources /honest-accounts-2017-how-world-profits-africas-wealth.

Godfrey, Ilan. *Legacy of the Mine*. Johannesburg: Jacana, 2013.

Gordimer, Nadine. "The Witwatersrand: A Time and Tailings." In *On the Mines*, edited by David Goldblatt and Nadine Gordimer, 19–26. Göttingen: Steidl, 2012.

Grace, Joshua. *African Motors: Technology, Gender, and the History of Development in Tanzania*. Durham, NC: Duke University Press, 2021.

Grace, Joshua. "Poop." *Somatosphere*, November 20, 2017.

Green, Lesley. *Rock/Water/Life: Ecology and Humanities for a Decolonial South Africa*. Durham, NC: Duke University Press, 2020.

Guinier, Lani, and Gerald Torres. *The Miner's Canary: Enlisting Race, Resisting Power, Transforming Democracy*. Cambridge, MA: Harvard University Press, 2003.

Guy, Jeff, and Motlatsi Thabane. "Technology, Ethnicity and Ideology: Basotho Miners and Shaft-Sinking on the South African Gold Mines." *Journal of Southern African Studies* 14, no. 2 (January 1988): 257–78.

Haraway, Donna J. *Staying with the Trouble: Making Kin in the Chthulucene*. Durham, NC: Duke University Press, 2016.

Harber, Jesse, and Kate Joseph. "Institutionalising the Gauteng City-Region." Governing the Gauteng City-Region Provocations no. 3. Johannesburg: GCRO, 2018.

Harrison, Philip, Graeme Götz, A. Todes, and Chris Wray, eds. *Changing Space, Changing City: Johannesburg after Apartheid*. Johannesburg: Wits University Press, 2014.

Hart, K. "On the Informal Economy: The Political History of an Ethnographic Concept." CEB Working Paper 09/042. Brussels: Centre Emile Bernheim, Université Libre de Bruxelles, 2009.

Haufler, Virginia. "Disclosure as Governance: The Extractive Industries Transparency Initiative and Resource Management in the Developing World." *Global Environmental Politics* 10, no. 3 (August 2010): 53–73.

Hecht, Gabrielle. "The African Anthropocene." *Aeon*, February 6, 2018.

Hecht, Gabrielle. *Being Nuclear: Africans and the Global Uranium Trade*. Cambridge, MA: MIT Press; Johannesburg: Wits University Press, 2012.

Hecht, Gabrielle. "Does Chernobyl Still Matter?" *Public Books*, November 22, 2019.

Hecht, Gabrielle. "Human Crap." *Aeon*, March 25, 2020.

Hecht, Gabrielle. "Interscalar Vehicles for an African Anthropocene: On Waste, Temporality, and Violence." *Cultural Anthropology* 33, no. 1 (February 22, 2018): 109–41.

Hecht, Gabrielle. *The Radiance of France: Nuclear Power and National Identity after World War II*. Cambridge, MA: MIT Press, 1998.

Hecht, Gabrielle. "Residue." *Somatosphere*, January 8, 2018.

Hepler-Smith, Evan. "Molecular Bureaucracy: Toxicological Information and Environmental Protection." *Environmental History* 24, no. 3 (July 2019): 534–60.

Herwitz, Daniel. *Heritage, Culture, and Politics in the Postcolony*. New York: Columbia University Press, 2012.

Hickel, Jason, Christian Dorninger, Hanspeter Wieland, and Intan Suwandi. "Imperialist Appropriation in the World Economy: Drain from the Global South through Unequal Exchange, 1990–2015." *Global Environmental Change* 73 (March 2022). https://doi.org/10.1016/j.gloenvcha.2022.102467.

Hinton, Elizabeth Kai. *America on Fire: The Untold History of Police Violence and Black Rebellion since the 1960s*. New York: Liveright, 2021.

Hoag, Colin. "'Water Is a Gift That Destroys': Making a National Natural Resource in Lesotho." *Economic Anthropology* 6, no. 2 (March 25, 2019): 183–94. https://doi.org/10.1002/sea2.12149.

Honwana, Alcinda Manuel. *Youth and Revolution in Tunisia*. London: Zed, 2013.

Hudson, Peter James. *Bankers and Empire: How Wall Street Colonized the Caribbean*. Chicago: University of Chicago Press, 2017.

Hufty, Marc. "Abandoned Mines: The Scars of the Past." *Global Challenges—Endangered Earth*, November 6, 2019.

Humby, Tracy-Lynn. "Environmental Justice and Human Rights on the Mining Wastelands of the Witwatersrand Gold Fields." *Revue Générale de Droit* 43 (January 13, 2014): 67–112.

Humby, Tracy-Lynn. "Facilitating Dereliction? How the South African Legal Regulatory Framework Enables Mining Companies to Circumvent Closure Duties." Ninth International Conference on Mine Closure, Sandton, South Africa, October 1–3, 2014.

Humby, Tracy-Lynn. "'One Environmental System': Aligning the Laws on the Environmental Management of Mining in South Africa." *Journal of Energy and Natural Resources Law* 33, no. 2 (April 3, 2015): 110–30.

Humby, Tracy-Lynn. "The Spectre of Perpetuity: Liability for Treating Acid Water on South Africa's Goldfields: Decision in Harmony II." *Journal of Energy and Natural Resources Law* 31, no. 4 (2013): 453–66.

Iheka, Cajetan. *African Ecomedia: Network Forms, Planetary Politics*. Durham, NC: Duke University Press, 2021.

International Human Rights Clinic. "The Cost of Gold: Environmental, Health, and Human Rights Consequences of Gold Mining in South Africa's West and Central Rand." Harvard Law School, October 2016.

International Network for Acid Prevention. "GARD Guide." Accessed October 8, 2021. http://www.gardguide.com/index.php?title=Main_Page.

IWQS (Institute for Water Quality Studies). "Report on the Radioactivity Monitoring Programme in the Mooi River (Wonderfonteinspruit) Catchment." Report No. N/C200/00//RPQ/2399. Department of Water Affairs and Forestry, 1999.

Jacobs, Jane. *Cities and the Wealth of Nations: Principles of Economic Life*. New York: Random House, 1984.

Jansen, Candice. "Colored Black: The Life and Works of South African Photographers Ernest Cole and Cedric Nunn." PhD diss., University of the Witwatersrand, 2020.

Jansson, Eva-Lotta. *An Acid River Runs through It*. Blurb, 2015.

Jenkins, Destin. *The Bonds of Inequality: Debt and the Making of the American City*. Chicago: University of Chicago Press, 2021.

Jenkins, Destin, and Justin Leroy, eds. *Histories of Racial Capitalism*. New York: Columbia University Press, 2021.

Johnson, Anthony. "Post-apartheid Citizenship and the Politics of Evictions in Inner City Johannesburg." PhD diss., The Graduate Center, City University of New York, 2016.

Johnson, Willard R. "Getting Over by Reaching Out: Lessons from the Divestment and Krugerrand Campaigns." *Black Scholar* 29, no. 1 (March 1999): 2–19.

Jordaan, J. M., J. F. Enslin, A. Havemann, L. E. Kent, and W. H. Cable. "Final Report of the Interdepartmental Committee on Dolomitic Mine Water: Far West Rand." Pretoria: Department of Water Affairs and Forestry, 1960.

"The June 16 Soweto Youth Uprising." South African History Online, 2013. https://www.sahistory.org.za/article/june-16-soweto-youth-uprising.

Katz, Elaine. *The White Death: Silicosis on the Witwatersrand Gold Mines, 1886–1910*. Johannesburg: Wits University Press, 1994.

Khumalo, Vusumuzi. "Kagiso Historical Research Report." Accessed May 19, 2020. https://www.mogalecity.gov.za/wp-content/uploads/Pdfs/kagiso_history.pdf?_t=1547619053.

Kisting, Sophia, Eugene Cairncross, David van Wyk, and Mariette Liefferink. "Case Study on Extractive Industries Prepared for the Lancet Commission on Global Governance: Report from South Africa." 2013. http://dx.doi.org/10.13140/RG.2.2.33297.45922.

Knape, Gunilla. "Notes on the Life of Ernest Cole." South African History Online, n.d. https://www.sahistory.org.za/sites/default/files/ERNEST%20COLE-gunilla.pdf.

Knowles, Scott Gabriel. *The Disaster Experts: Mastering Risk in Modern America*. Philadelphia: University of Pennsylvania Press, 2011.

Knowles, Scott Gabriel. "Learning from Disaster? The History of Technology and the Future of Disaster Research." *Technology and Culture* 55, no. 4 (December 3, 2014): 773–84. https://doi.org/10.1353/tech.2014.0110.

Knowles, Scott Gabriel. "Slow Disaster in the Anthropocene: A Historian Witnesses Climate Change on the Korean Peninsula." *Daedalus* 149, no. 4 (October 1, 2020): 192–206.

Kojola, Erik, and David N. Pellow. "New Directions in Environmental Justice Studies: Examining the State and Violence." *Environmental Politics* 30, nos. 1–2 (February 23, 2021): 100–118.

Kootbodien, T., A. Mathee, N. Naicker, and N. Moodley. "Heavy Metal Contamination in a School Vegetable Garden in Johannesburg." *South African Medical Journal* 102, no. 4 (April 2012): 226–27.

Krupar, Shiloh. "Brownfields as Waste/Race Governance: US Contaminated Property Redevelopment and Racial Capitalism." In *The Routledge Handbook of Waste Studies*, edited by Zsuzsa Gille and Josh Lepawsky, 238–53. London: Routledge, 2022.

Kuklick, Henrika. "Contested Monuments: The Politics of Archeology in Southern Africa." In *Colonial Situations: Essays on the Contextualization of Ethnographic Knowledge*, edited by George W. Stocking Jr., 135–69. Madison: University of Wisconsin Press, 1991.

Larkin, Jason. *Tales from the City of Gold*. Berlin: Kehrer Verlag, 2013.

Ledwaba, Lucas, and Leon Sadiki. *Broke and Broken: The Shameful Legacy of Gold Mining in South Africa*. Auckland Park, South Africa: Blackbird, 2016.

Legassick, Martin, and David Hemson. "Foreign Investment and the Reproduction of Racial Capitalism in South Africa." Anti-Apartheid Movement, September 1976. https://www.sahistory.org.za/sites/default/files/archive_files/Foreign%20Investment%20by%20%20Martin%20Legassick%20%26%20Dave%20Hemson.pdf.

le Roux, Hannah. "Designing KwaThema: Cultural Inscriptions in the Model Township." *Journal of Southern African Studies* 45, no. 2 (2019): 273–307.

le Roux, Hannah, and Gabrielle Hecht. "Bad Earth." *e-Flux Architecture*, August 31, 2020.

Levin, Kelly, Benjamin Cashore, Steven Bernstein, and Graeme Auld. "Overcoming the Tragedy of Super Wicked Problems: Constraining Our Future Selves to Ameliorate Global Climate Change." *Policy Sciences* 45, no. 2 (June 2012): 123–52.

Lewis, Mark, and Tanya Zack. *Undercity: Wake Up, This Is Joburg, Vol. 10*. Johannesburg: Fourthwall, 2018.

Liboiron, Max. *Pollution Is Colonialism*. Durham, NC: Duke University Press, 2021.

Liboiron, Max, Manuel Tironi, and Nerea Calvillo. "Toxic Politics: Acting in a Permanently Polluted World." *Social Studies of Science* 48, no. 3 (2018): 331–49.

Lichtenstein, Alex. "Mining Nostalgia." *Los Angeles Review of Books*, July 22, 2019.

Lichtenstein, Alex, and Rick Halpern. *Margaret Bourke-White and the Dawn of Apartheid*. Bloomington: Indiana University Press, 2016.

Liefferink, Mariette. "Assessing the Past and Present Role of the National Nuclear Regulator as a Public Protector against Potential Health Injuries: The West and Far West Rand Case Study." Unpublished manuscript, 2012.

Liefferink, Mariette. "Classification of FSE's Activism Pertaining to AMD." Unpublished manuscript, 2011.

Livingston, Julie. *Debility and the Moral Imagination in Botswana*. Bloomington: Indiana University Press, 2005.

Livingston, Julie. *Self-Devouring Growth: A Planetary Parable as Told from Southern Africa*. Durham, NC: Duke University Press, 2019.

Lombard, Marlize. *On the Minds of Bow Hunters: Squeezing Minds from Stones*. Oxford: Oxford University Press.

Lyssens, Marjolein, and Griet Juwet. "Sculpting the Hill, Cultivating the Valley: Re-claiming the West Rand's Post-mining Landscape." Master in Urbanism and Strategic Planning thesis, Katholieke Universiteit Leuven, 2014.

Madida, Sipokazi. "Exploring the Conflicting Meanings of the South African Post-apartheid Heritage Practice." Online talk, May 12, 2021, Stanford University.

Madihlaba, Thabo. "The Environmental Impact of Mining on Communities in South Africa." In *Environmental Justice in South Africa*, edited by David McDonald, 156–67. Athens: Ohio University Press, 2002.

Madonsela, Sanet. "Critical Reflections on State Capture in South Africa." *Insight on Africa* 11, no. 1 (January 2019): 113–30.

Mamdani, Mahmood. *Neither Settler nor Native: The Making and Unmaking of Permanent Minorities*. Cambridge, MA: Belknap, 2020.

Maneli, Boyce Makhosonke, and Austin Musundire. "Investigating the Impact of the Integrated Approach on Local Economic Development in South Africa in the Context of Policy Planning and Implementation: The Case of the West Rand District Municipality and Its Development Agency." *International Journal of Science and Research* 9, no. 7 (2020): 10.

Manungufala, T. E., M. Sabiti-Kalule, and I. Aucamp. "Investigation into the Slimes Dams, Mine Dumps and Landfills (Residue Deposits) as Environmental Constraints to Low-Cost Housing Projects in Gauteng, South Africa." Twenty-Third IAHS World Congress on Housing, Pretoria, South Africa, 2005.

Martin, Michaela Elsbeth, and Hussein Solomon. "Understanding the Phenomenon of 'State Capture' in South Africa." *Southern African Peace and Security Studies* 5, no. 2 (2016): 139–69.

Marutlulle, Noah K. "A Critical Analysis of Housing Inadequacy in South Africa and Its Ramifications." *Africa's Public Service Delivery and Performance Review* 9, no. 1 (March 24, 2021): 16.

Masuku, Themba, Malose Langa, and David Bruce. "The Community Work Programme in Kagiso." Centre for the Study of Violence and Reconciliation, April 2016.

Mavhunga, Clapperton Chakanetsa. *The Mobile Workshop: The Tsetse Fly and African Knowledge Production*. Cambridge, MA: MIT Press, 2018.

Mavhunga, Clapperton Chakanetsa. *Transient Workspaces: Technologies of Everyday Innovation in Zimbabwe*. Cambridge, MA: MIT Press, 2014.

Mavhunga, Clapperton Chakanetsa, ed. *What Do Science, Technology, and Innovation Mean from Africa?* Cambridge, MA: MIT Press, 2017.

Maviyane-Davies, Chaz. *A World of Questions: 120 Posters on the Human Condition*. Mzingeli, 2015.

Mbembe, Achille. *Critique of Black Reason*. Translated by Laurent Dubois. Durham, NC: Duke University Press, 2017.

Mbembe, Achille. *Out of the Black Night: Essays on Decolonization*. New York: Columbia University Press, 2021.

Mbembe, Achille, and Carolyn Shread. "The Universal Right to Breathe." *Critical Inquiry* 47, no. S2 (December 22, 2020): S58–S62.

McCarthy, T. S., G. Steyl, J. Maree, and B. Zhao. "Mine Water Management in the Witwatersrand Gold Fields with Special Emphasis on Acid Mine Drainage." Report to the Inter-Ministerial Committee on Acid Mine Drainage. December 2010. https://www.dmr.gov.za/Portals/0/Resource%20Center/Reports%20and%20Other%20Documents/2010_Report_Interministerial%20Committee_Acid%20Mine%20Drainage.pdf?ver=2018-03-13-020432-273.

McCarthy, Terence. "The Decanting of Acid Mine Water in the Gauteng City-Region: Analysis, Prognosis and Solutions." Governing the Gauteng City-Region Provocations no. 1. Johannesburg: GCRO, 2010.

McCulloch, Jock. *Asbestos Blues: Labour, Capital, Physicians and the State in South Africa*. Oxford: James Currey, 2002.

McCulloch, Jock. "Asbestos, Lies and the State: Occupational Disease and South African Science." *African Studies*, no. 2 (2005): 201–16.

McCulloch, Jock. "Mining Evidence: South Africa's Gold Mines and the Career of A. J. Orenstein." *Social History of Medicine* 31, no. 1 (February 1, 2018): 61–78.

McCulloch, Jock. *South Africa's Gold Mines and the Politics of Silicosis*. Woodbridge, UK: James Currey, 2012.

McDonald, David A., ed. *Environmental Justice in South Africa*. Athens: Ohio University Press, 2002.

McGill, R. A., R. E. Fox, and A. R. Hughes. "Rehabilitation of the Nabarlek Uranium Mine—Will Close Out Ever Be Achieved?" In *Uranium in the Aquatic Environment*, edited by Broder J. Merkel, Britta Planer-Friedrich, and Christian Wolkersdorfer, 727–36. Berlin: Springer, 2002.

Mendes, Margarida. "Molecular Colonialism." Anthropocene Campus 2016: Techno-Metabolism, May 29, 2017. https://www.anthropocene-curriculum.org/contribution/molecular-colonialism.

Merkel, Broder J., Britta Planer-Friedrich, and Christian Wolkersdorfer, eds. *Uranium in the Aquatic Environment*. Berlin: Springer, 2002.

Meskell, Lynn. *The Nature of Heritage: The New South Africa*. Chichester: Wiley-Blackwell, 2011.

Milaras, M., F. Ahmed, and T. J. M. McKay. "Mine Closure in South Africa: A Survey of Current Professional Thinking and Practice." Johannesburg: University of the Witwatersrand, 2014.

Millar, Kathleen M. "Garbage as Racialization." *Anthropology and Humanism* 45, no. 1 (June 1, 2020): 4–24.

Miller, Duncan, Nirdev Desai, and Julia Lee-Thorp. "Indigenous Gold Mining in Southern Africa: A Review." *Goodwin Series* 8 (2000): 91–99. https://doi.org/10.2307/3858050.

Mills, Charles W. "Black Trash." In *Faces of Environmental Racism: Confronting Issues of Global Justice*, 2nd ed., edited by Laura Westra and Bill E. Lawson, 73–91. Lanham, MD: Rowman and Littlefield, 2001.

Mills, Charles W. *The Racial Contract*. Ithaca, NY: Cornell University Press, 1997.

Mkhize, Thembani. "Rescaling Municipal Governance in Gauteng: Competing Rationalities in Sedibeng's Proposed Re-demarcation and Metropolitanisation." *Urban Forum* 32, no. 2 (June 2021): 205–23.

Mnwana, Sonwabile. "'Custom' and Fractured 'Community': Mining, Property Disputes and Law on the Platinum Belt, South Africa." *Third World Thematics* 1, no. 2 (2016): 218–34.

Mnwana, Sonwabile. "Democracy, Development and Chieftaincy along South Africa's 'Platinum Highway': Some Emerging Issues." *Journal of Contemporary African Studies* 33, no. 4 (October 2015): 510–29.

Mnwana, Sonwabile. "A Radical Mineral Policy with Retribalisation? Mining and Politics of Difference in Rural South Africa." In *Extractive Industries and Changing State Dynamics in Africa: Beyond the Resource Curse*, edited by Jon Schubert, Ulf Engel, and Elísio Salvado Macamo, 112–28. New York: Routledge, 2018.

Mnwana, Sonwabile, and Andrew Bowman. "Land, Conflict and Radical Distributive Claims in South Africa's Rural Mining Frontier." *Extractive Industries and Society* 11 (2022). https://doi.org/10.1016/j.exis.2021.100972.

Molelekwa, Thabo. "Silicosis Payouts Are 'Symbolic Justice' for Miners." *New Frame*, September 13, 2021. https://www.newframe.com/silicosis-payouts-are-symbolic-justice-for-miners/.

Moloi, Dudley. "When Gold Becomes a Nuisance." In *Environmental Justice in South Africa*, edited by David McDonald, 168–70. Athens: Ohio University Press, 2002.

Mondi, Lumkile Patriarch, and Simon Roberts. "The Role of Development Finance for Industry in a Restructuring Economy: A Critical Reflection on the Industrial Development Corporation of South Africa." United Nations University, 2005.

Moore, Jason W. *Capitalism in the Web of Life: Ecology and the Accumulation of Capital*. New York: Verso, 2015.

Moore, Rob. "Taking the Gap: Reflections on Hybridity." Johannesburg: GCRO, May 31, 2021. https://www.gcro.ac.za/news-events/news/detail/outgoing-ed-rob-moores-farewell-message/.

Morris, Rosalind. "Death and the Miner." *Berlin Journal*, October 9, 2018.

Morris, Rosalind. "Shadow and Impress: Ethnography, Film and the Task of Writing History in the Space of South Africa's Deindustrialization." *History and Theory* 57, no. 4 (December 1, 2018): 102.

Moshupya, Paballo, Tamiru Abiye, Hassina Mouri, Mannie Levin, Marius Strauss, and Rian Strydom. "Assessment of Radon Concentration and Impact on Human Health in a Region Dominated by Abandoned Gold Mine Tailings Dams: A Case from the West Rand Region, South Africa." *Geosciences* 9, no. 11 (November 2019). https://doi.org/10.3390/geosciences9110466.

Mpofu-Walsh, Sipho. *The New Apartheid*. Cape Town: Tafelberg, 2021.

Mtolo, Khanyiso Edmund. "Effectiveness of Environmental Management Frameworks in South Africa: Evaluating Stakeholder Perceptions and Expectations." Master of Environment and Development thesis, University of KwaZulu-Natal, 2010.

Mudd, Gavin M. "Global Trends in Gold Mining: Towards Quantifying Environmental and Resource Sustainability." *Resources Policy* 32, nos. 1–2 (2007): 42–56. https://doi.org/10.1016/j.resourpol.2007.05.002.

Mudd, Gavin M. "The Limits to Growth and 'Finite' Mineral Resources: Re-visiting the Assumptions and Drinking from That Half-Capacity Glass." *International Journal of Sustainable Development* 16, nos. 3–4 (2013): 204–20.

Mujere, Joseph. "Unemployment, Service Delivery and Practices of Waiting in South Africa's Informal Settlements." *Critical African Studies* 12, no. 1 (January 2, 2020): 65–78.

Murphy, Michelle. *The Economization of Life*. Durham, NC: Duke University Press, 2017.

Murray, Martin J. *City of Extremes: The Spatial Politics of Johannesburg*. Durham, NC: Duke University Press, 2011.

Murray, Martin J. *Panic City: Crime and the Fear Industries in Johannesburg*. Stanford, CA: Stanford University Press, 2020.

Mushai, Albert. "The Long Road to Compensation for Silicosis Sufferers in South Africa." *Journal of Southern African Studies* 46, no. 6 (November 1, 2020): 1127–43.

Ndlovu, Sifiso Mxolisi. *The Soweto Uprisings: Counter Memories of June 1976*. Johannesburg: Ravan, 1998.

Nelson, Alondra. *Body and Soul: The Black Panther Party and the Fight against Medical Discrimination*. Minneapolis: University of Minnesota Press, 2011.

Ngole-Jeme, Veronica Mpode, and Peter Fantke. "Ecological and Human Health Risks Associated with Abandoned Gold Mine Tailings Contaminated Soil." *PLOS ONE* 12, no. 2 (February 21, 2017). https://doi.org/10.1371/journal.pone.0172517.

Nieftagodien, Noor. *The Soweto Uprising*. Athens: Ohio University Press, 2014.

Nieftagodien, Noor, and Sally Gaule. *Orlando West, Soweto: An Illustrated History*. Johannesburg: Wits University Press, 2012.

Niekerk, H. J. Van, and M. J. Viljoen. "Causes and Consequences of the Mer-riespruit and Other Tailings-Dam Failures." *Land Degradation and Development* 16, no. 2 (2005): 201–12.

"1946 African Mineworkers Strike." South African History Online, accessed February 2, 2022. https://www.sahistory.org.za/article/1946-african-mineworkers-strike.

NNR (National Nuclear Regulator). "Regulatory Strategy for the Remediation of the Wonderfontein Catchment Area." NNR-SD-0002. National Nuclear Regulator, 2009.

Nørgård, Jørgen Stig, John Peet, and Kristín Vala Ragnarsdóttir. "The History of *The Limits to Growth*." *Solutions Journal*, February 22, 2016.

Nuttall, Sarah, and Achille Mbembe. *Johannesburg: The Elusive Metropolis*. Durham, NC: Duke University Press, 2008.

Oesi, Joseph, dir. *Black Lives Matter*. South Africa, 2016.

Olalde, Mark. "The Haunting Legacy of South Africa's Gold Mines." *Yale E360*, November 12, 2015.

Oreskes, Naomi, and Erik M. Conway. *Merchants of Doubt: How a Handful of Scientists Obscured the Truth on Issues from Tobacco Smoke to Global Warming*. New York: Bloomsbury, 2010.

Packard, Randall M. *White Plague, Black Labor: Tuberculosis and the Political Economy of Health and Disease in South Africa*. Berkeley: University of California Press, 1989.

Park, Emma, and Kevin P. Donovan. "Between the Nation and the State." *Limn*, August 9, 2016. https://limn.it/articles/between-the-nation-and-the-state/.

Parker, Alexandra, Christian Hamann, and Julia de Kadt. "Accessing Quality Education in Gauteng: Intersecting Scales of Geography, Educational Policy and Inequality." *Urban Forum* 32, no. 2 (June 2021): 141–63.

Parliamentary Monitoring Group. "National Nuclear Regulator on Relocation of Tudor Shaft Informal Settlement from Mine Dump Contaminated with Nuclear Radioactive Material." February 28, 2011. https://pmg.org.za/committee-meeting/12639/.

Parvin, Nassim, and Anne Pollock. "Unintended by Design: On the Political Uses of 'Unintended Consequences.'" *Engaging Science, Technology, and Society* 6 (August 1, 2020): 320–27.

Pellow, David Naguib. *Resisting Global Toxics: Transnational Movements for Environmental Justice*. Cambridge, MA: MIT Press, 2007.

Pellow, David Naguib, and Robert J. Brulle, eds. *Power, Justice, and the Environment: A Critical Appraisal of the Environmental Justice Movement*. Cambridge, MA: MIT Press, 2005.

Peterson, Derek R., Kodzo Gavua, and Ciraj Rassool, eds. *The Politics of Heritage in Africa: Economies, Histories, and Infrastructures*. Ann Arbor: University of Michigan Press, 2015.

Phasha, Potšišo. *A City on a Hill*. Self-published: Johannesburg, 2021.

Phillips, O. "Surveillance Report of the Upper Wonderfonteinspruit Catchment Area." National Nuclear Regulator, March 14, 2011.

Phumla Patience Mjadu (and 291 further residents of the Tudor Shaft Informal Settlement) and Federation for a Sustainable Environment v The National Nuclear Regulator and Mogale City Local Municipality, No. 24611/2012, High Court of South Africa (South Gauteng, Johannesburg Division) 2012.

Pigou, Arthur Cecil. *The Economics of Welfare*. London: Palgrave Macmillan, 2013.

Posel, Deborah. "The Case for a Welfare State: Poverty and the Politics of the Urban African Family in the 1930s and 1940s." In *South Africa's 1940s: World of Possibilities*, edited by Saul Dubow and Alan Jeeves, 64–86. Cape Town: Double Storey, 2005.

Power, Michael. *The Audit Society*. Oxford: Oxford University Press, 1997.

Pretorius, Theo. "West Mining Belt: Formulation of a Strategic Area Framework. Prepared for City of Johannesburg Metropolitan Municipality." Plan Associates, in association with Novus, Iyer Urban Design Studio, Gudani Consulting, GLS Consulting, April 2016.

Proctor, Robert. *Golden Holocaust: Origins of the Cigarette Catastrophe and the Case for Abolition*. Berkeley: University of California Press, 2012.

Proctor, Robert, and Londa L. Schiebinger. *Agnotology: The Making and Unmaking of Ignorance*. Stanford, CA: Stanford University Press, 2008.

Pulido, Laura. "Flint, Environmental Racism, and Racial Capitalism." *Capitalism Nature Socialism* 27, no. 3 (July 2, 2016): 1–16.

Pulles, W. "Radionuclides in South African Gold Mining Water Circuits: An Assessment of Licensing, Health Hazards, Water and Waste Water Regulations and Impact on the Environment and Workforce." Restricted Report No. 17/91. COMRO, 1991.

Rasmussen, Jacob. "Struggling for the City: Evictions in Inner-City Johannesburg." In *The Security-Development Nexus: Expressions of Sovereignty and Securitization in Southern Africa*, edited by Lars Buur, Steffen Jensen, and Finn Stepputat, 174–90. Uppsala: Nordiska Afrikainstitute; Cape Town: HSRC Press, 2007.

Rassool, Ciraj. "The Rise of Heritage and the Reconstitution of History in South Africa." *Kronos*, no. 26 (2000): 1–21.

Rassool, Ciraj, and Leslie Witz. "South Africa: A World in One Country. Moments in International Tourist Encounters with Wildlife, the Primitive and the Modern." *Cahiers d'Études Africaines* 36, no. 143 (1996): 335–71.

Redfield, Peter, and Steven Robins. "An Index of Waste: Humanitarian Design, 'Dignified Living' and the Politics of Infrastructure in Cape Town." *Anthropology Southern Africa* 39, no. 2 (May 31, 2016): 145–62.

Reichardt, Markus. "The Wasted Years: A History of Mine Waste Rehabilitation Methodology in the South African Mining Industry from Its Origins to 1991." PhD diss., University of the Witwatersrand, 2013.

Reid, Jasmine. "Dwelling in Displacement: Land Rights and Heritage Activism in Post-apartheid Johannesburg." PhD diss., Stanford University, 2022.

Reno, Joshua A., and Britt Halvorson. "Waste and Whiteness." In *The Routledge Handbook of Waste Studies*, edited by Zsuzsa Gille and Josh Lepawsky, 41–54. London: Routledge, 2022.

Reynard, Jerome, Justin Bradfield, Marlize Lombard, and Sarah Wurz. "What a Bone Arrowhead from South Africa Reveals about Ancient Human Cognition." *The Conversation*, May 17, 2020. http://theconversation.com/what-a-bone-arrowhead-from-south-africa-reveals-about-ancient-human-cognition-137651.

Rittel, Horst W. J., and Melvin M. Webber. "Dilemmas in a General Theory of Planning." *Policy Sciences* 4, no. 2 (June 1, 1973): 155–69.

Rodney, Walter. *How Europe Underdeveloped Africa*. London: Verso, 2018.

Samson, Melanie. "Accumulation by Dispossession and the Informal Economy—Struggles over Knowledge, Being and Waste at a Soweto Garbage Dump." *Environment and Planning D: Society and Space* 33, no. 5 (October 1, 2015): 813–30.

Schadeberg, Jurgen, dir. "Ernest Cole." Journeyman TV, 2006.

Schrader, Aljoscha, and Frank Winde. "Unearthing a Hidden Treasure: 60 Years of Karst Research in the Far West Rand, South Africa." *South African Journal of Science* 111, nos. 5–6 (May 28, 2015). https://doi.org/10.17159/sajs.2015/20140144.

Scott, Allen J., ed. *Global City-Regions: Trends, Theory, Policy*. Oxford: Oxford University Press, 2001.

Sello, Kefiloe. "Rivers That Become Reservoirs: An Ethnography of Water Commodification in Lesotho." PhD diss., University of Cape Town, 2022.

Sherman, Richard. "Forgotten People, Episode 1: Mogale City." 2012. https://www.youtube.com/watch?v=paLB2teh6j0&t=2s.

Shivamba, Amanda. "Mining for Sustainable Development Research Report." Corruption Watch, 2017.

Smil, Vaclav. *Growth: From Microorganisms to Megacities*. Cambridge, MA: MIT Press, 2020.

Smith, Tymon. "The Mystery of Ernest Cole's Archive." Parts 1 and 2, *New Frame*, February 20, 2020. https://www.newframe.com/part-one-the-mystery-of-ernest-coles-archive/.

South African Human Rights Commission. "National Hearing on the Underlying Socio-economic Challenges of Mining-Affected Communities in South Africa." Johannesburg: South African Human Rights Commission, 2016.

Sparks, Stephen. "Between 'Artificial Economics' and the 'Discipline of the Market': Sasol from Parastatal to Privatisation." *Journal of Southern African Studies* 42, no. 4 (2016): 711–24.

Sparks, Stephen. "Crude Politics: The African National Congress, the Shipping Research Bureau and the Anti-apartheid Oil Boycott, c1960–1979." *South African Historical Journal* 69, no. 2 (2017): 251–64.

Stasik, Michael, Valerie Hänsch, and Daniel Mains. "Temporalities of Waiting in Africa." *Critical African Studies* 12, no. 1 (January 2, 2020): 1–9.

Stoch, L. E. J. "Krugersdorp Game Reserve: Animal Mortality." Unpublished memorandum on behalf of the Council for Geosciences, 2005.

Stoch, L. E. J. "Some Controversial Health Related Issues in a Karst Environment." Unpublished manuscript, August 1, 2006.

Stoch, L. E. J., F. Winde, and Ewald Erasmus. "Karst, Mining and Conflict: A Historical Perspective of the Consequences of Mining on the Far West Rand." Unpublished manuscript, n.d.

Strathern, Marilyn, ed. *Audit Cultures: Anthropological Studies in Accountability, Ethics and the Academy*. London: Routledge, 2000.

Tang, Dorothy, and Andrew Watkins. "Ecologies of Gold." *Places Journal*, February 24, 2011.

Taylor, Dorceta E. *The Environment and the People in American Cities, 1600s–1900s: Disorder, Inequality, and Social Change*. Durham, NC: Duke University Press, 2009.

Taylor, Dorceta E. *The Rise of the American Conservation Movement: Power, Privilege, and Environmental Protection*. Durham, NC: Duke University Press, 2016.

Thain, G. M. "Ancient Gold-Mining Sites in the Transvaal." *Journal of the South African Institute of Mining and Metallurgy*, January 1974, 243–44.

Thomas, Julia Adeney. "Why the Anthropocene Is Not 'Climate Change'—and Why That Matters." *Climate and Capitalism*, January 31, 2019.

Thornton, Robert. "Zamazama, 'Illegal' Artisanal Miners, Misrepresented by the South African Press and Government." *Extractive Industries and Society* 1, no. 2 (November 2014): 127–29.

Toens, P. D., W. Stadler, and N. J. Wullschlege. "The Association of Groundwater Chemistry and Geology with Atypical Lymphocytes (as Biological Indicator) in the Pofadder Area, North Western Cape, South Africa." WRC report 839/1/98. 1998.

Toffah, T. "Mines of Gold, Mounds of Dust: Resurrecting the Witwatersrand." Master's thesis, Katholieke Universiteit Leuven, 2012.

Tousignant, Noémi. *Edges of Exposure: Toxicology and the Problem of Capacity in Postcolonial Senegal*. Durham, NC: Duke University Press, 2018.

Trangos, Guy, and Kerry Bobbins. "Gold Mining Exploits and the Legacies of Johannesburg's Mining Landscapes." *Scenario Journal* (blog), October 8, 2015. http://scenariojournal.com/article/gold-mining-exploits/.

Tsing, Anna L., Jennifer Deger, Alder Keleman Saxena, and Feifei Zhou, eds. *Feral Atlas: The More-Than-Human Anthropocene*, Stanford, CA: Stanford University Press, 2020. http://doi.org/10.21627/2020fa.

Turton, Anthony, Craig Schultz, Hannes Buckle, Mapule Kgomongoe, Tinyiko Malungani, and Mikael Drackner. "Gold, Scorched Earth and Water: The Hydropolitics of Johannesburg." *International Journal of Water Resources Development* 22, no. 2 (June 2006): 313–35.

United Nations Environment Programme. "Abandoned Mines: Problems, Issues and Policy Challenges for Decision Makers. Santiago, Chile, 18 June 2001: Summary Report." Santiago: Chilean Copper Commission and United Nations Environment Programme, 2001.

Valentin, J., ed. *The 2007 Recommendations of the International Commission on Radiological Protection*. Annals of the ICRP, Publication 103. Amsterdam: Elsevier, 2007.

van Heerden, Carina. "The Death of a Mine Dump." *Dekat*, February 2010, 62–69.

Vergès, Françoise. "Capitalocene, Waste, Race, and Gender." *e-Flux* 100 (May 2019).

Verwoerd, Hendrik Frensch. "Policy of the Minister of Native Affairs, May 30, 1952." In *Verwoerd Speaks; Speeches 1948–1966*, edited by A. N. Pelzer. Johannesburg: APD, 1966.

Vladislavić, Ivan. *Portrait with Keys: The City of Johannesburg Unlocked*. London: Portobello, 2006.

Vladislavić, Ivan. *The Restless Supermarket*. Cape Town: David Philip, 2001.

von Schnitzler, Antina. *Democracy's Infrastructure: Techno-Politics and Protest after Apartheid*. Princeton, NJ: Princeton University Press, 2016.

Voyles, Traci Brynne. *Wastelanding: Legacies of Uranium Mining in Navajo Country*. Minneapolis: University of Minnesota Press, 2015.

Wade, P. W., S. Woodborne, W. M. Morris, P. Vos, and N. W. Jarvis. "Tier 1 Risk Assessment of Selected Radionuclides in Sediments of the Mooi River Catchment." Water Research Commission, 2002.

Waggitt, Peter. "Changing Standards and Continuous Improvement: A History of Uranium Mine Rehabilitation in Northern Australia." In *Uranium in the Aquatic Environment*, edited by Broder J. Merkel, Britta Planer-Friedrich, and Christian Wolkersdorfer, 645–52. Berlin: Springer, 2002.

Wambecq, W., T. Toffah, B. De Meulder, and H. le Roux. "From Mine to Life Cycle Closure: Towards a Vision for the West Rand, Johannesburg." In *Mine Closure 2014*. Johannesburg: University of the Witwatersrand, 2014.

Watson, I., and M. Olalde. "The State of Mine Closure in South Africa—What the Numbers Say." *Journal of the Southern African Institute of Mining and Metallurgy* 119, no. 7 (2019): 639–45.

Weizman, Eyal. *Hollow Land: Israel's Architecture of Occupation*. London: Verso, 2007.

Wendel, G. "Radioactivity in Mines and Mine Water—Sources and Mechanisms." *Journal of the South African Institute of Mining and Metallurgy* (March/April 1998): 87–92.

White, Richard. *The Organic Machine*. New York: Hill and Wang, 1995.

WHO (World Health Organization). "Uranium in Drinking Water: Summary Statement." In *WHO Guidelines for Drinking-Water Quality*, 3rd ed. WHO/SDE/WSH/03,04/118. Geneva: WHO, 2004.

Wilhelm-Solomon, Matthew. "Decoding Dispossession: Eviction and Urban Regeneration in Johannesburg's Dark Buildings." *Singapore Journal of Tropical Geography* 37, no. 3 (2016): 378–95.

Winde, Frank. "Questions Regarding the Scope of Work of the Requested Study on a Methodology for Remediation." Presentation to the Radiation Steering Committee, National Nuclear Regulator, Centurion, August 5, 2011.

Winde, Frank. "Uranium Pollution of Water: A Global Perspective on the Situation in South Africa." Vanderbijlpark: North-West University, Vaal Triangle Campus, 2013.

Winde, Frank. "Uranium Pollution of Water Resources in Mined-Out and Active Goldfields of South Africa: A Case Study of the Wonderfonteinspruit Catchment on Extent and Sources of U-Contamination and Associated Health Risks." In *Abstracts of the International Mine Water Conference*. Pretoria: IMWA, 2009.

Winde, Frank, and Abraham Barend de Villiers. "The Nature and Extent of Uranium Contamination from Tailings Dams in the Witwatersrand Gold Mining Area (South Africa)." In *Uranium in the Aquatic Environment*, edited by Broder J. Merkel, Britta Planer-Friedrich, and Christian Wolkersdorfer, 889–98. Berlin: Springer, 2002.

Winde, Frank, and Abraham Barend de Villiers. "Uranium Contamination of Streams by Tailings Deposits—Case Studies in the Witwatersrand Gold Mining Area (South Africa)." In *Uranium in the Aquatic Environment*, edited by Broder J. Merkel, Britta Planer-Friedrich, and Christian Wolkersdorfer, 803–12. Berlin: Springer, 2002.

Winde, Frank, Ewald Erasmus, and Gerhard Geipel. "Uranium Contaminated Drinking Water Linked to Leukaemia—Revisiting a Case Study from South Africa Taking Alternative Exposure Pathways into Account." *Science of the Total Environment* 574 (January 2017): 400–421.

Winde, Frank, Gerhard Geipel, Carolina Espina, and Joachim Schüz. "Human Exposure to Uranium in South African Gold Mining Areas Using Barber-Based Hair Sampling." *PLOS ONE* 14, no. 6 (June 27, 2019). https://doi.org/10.1371/journal.pone.0219059.

Winde, F[rank], E. Hoffmann, C. Espina, and J. Schüz. "Mapping and Modelling Human Exposure to Uraniferous Mine Waste Using a GIS-Supported Virtual Geographic Environment." *Journal of Geochemical Exploration* 204 (September 1, 2019): 167–80.

Winde, Frank, and Izak Jacobus van der Walt. "The Significance of Groundwater–Stream Interactions and Fluctuating Stream Chemistry on Waterborne Uranium Contamination of Streams—a Case Study from a Gold Mining Site in South Africa." *Journal of Hydrology* 287, nos. 1–4 (February 2004): 178–96.

Winterer, Caroline. *How the New World Became Old: The Deep Time Revolution in America*. Princeton, NJ: Princeton University Press, forthcoming.

Witz, Leslie. *Apartheid's Festival*. Bloomington: Indiana University Press, 2003.

Witz, Leslie. *Unsettled History: Making South African Public Pasts*. Ann Arbor: University of Michigan Press, 2017.

Witz, L[eslie], C. Rassool, and G. Minkley. "Repackaging the Past for South African Tourism." *Daedalus* 130 (2001): 277–96.

Woodson, Carter Godwin. *The Mis-education of the Negro*. Washington, DC: Associated Publishers, 1933.

Wright, Gwendolyn. *The Politics of Design in French Colonial Urbanism*. Chicago: University of Chicago Press, 1991.

Yusoff, Kathryn. *A Billion Black Anthropocenes or None*. Minneapolis: University of Minnesota Press, 2018.

Zalasiewicz, Jan, Colin N. Waters, Mark Williams, Anthony D. Barnosky, Alejandro Cearreta, Paul Crutzen, Erle Ellis, et al. "When Did the Anthropocene Begin? A Mid-Twentieth Century Boundary Level Is Stratigraphically Optimal." *Quaternary International* 383 (October 2015): 196–203.

Zimring, Carl A. *Clean and White: A History of Environmental Racism in the United States*. New York: NYU Press, 2017.

INDEX

Note: Page numbers in italics refer to illustrations.

abandoned mines, 25, 45, 66, 109, 164, 166; abandoned people and, 207; accountability and, 169–70; closure and rehabilitation, 186–90, 193–94; in South Africa, 29, *165*; worldwide, 199

Abrahams, Peter, 32, 109, 119; "Fancies Idle," 93, *98*

acid mine drainage (AMD), 7, 12, 45, 86; abandoned mines and, 25, 199; description of, 25, 49–50; GCR water supply and, 170; historical observations of, 51; international expertise and, 66–68; liability and remediation, 160–61, 169, 187, 188; mitigation costs, 171; in rivers and streams, 53, 64, *65*, 163; West Rand decant and treatment plant, *50*, *59*, 61, 63, 81–83, *83*, 159

African National Congress (ANC), 48, 72, 149, 165–66, 191; Community Work Program (CWP), 134–36, 227n15; elections, 129, 179–80; megaprojects, 179, 183; political rivalry, 151–52; residual governance and, 181. *See also* Reconstruction and Development Programme (RDP) homes

Afrikaans, 1–2, 4, 21, 47, 72, 133

Agricola, Georgius, 44, 51–52, 67

Alexander, Amanda, 165–66

Alexander, Neville, 10

Alexandra township, 60–61

Amatshe Mining, 160–61

Anglo American, 66, 93, 113

animals: heavy metals and, 54, 74; water pollution and, 70, 72

Anthropocene, 9, 11, 16, 29, 34, 41, 204; golden spike, 44; racial capitalism and, 5, 6, 12, 31, 32, 207; residual governance and, 198; views from space and, 20, 25; waste and, 27, 28, 199

anti-apartheid movement, 38, 42

apartheid, 7, 32, 41, 97; buffer zones, 107–8, 113, 164; Community Councils, 133; education policy, 1–2, 4; forced removals under, *107*, 142, *143*; infrastructures, 4, 5, 169–70; liberation struggles, 133, 205–6; mining industry and, 9, 33, 40, 190; photographic documentation, 90–93, *94–96*; privatization and, 69, 186; racial capitalism and, 10, 205; residual governance and, 138; secrecy and, 52; spatiality of, 25, 86, 102, 164, 180; traditional governance and, 191, 233n70

Apartheid Museum, *95*, 183

aquatic environments, *50*, 66, 67. *See also* fish

archaeology, 35, 37

arsenic, 49, 100, 138, 229n63

asbestos, 55, 78, 89, 108, 133; mines, 193

astrobleme, 35, 38, 42
atomic bombs, 44, 109
Atomic Energy Board (AEB), 44, 54
Australopithecus, 35, *36*

Ballard, Richard, 180, 181
Bantu, 102–3; education policy, 1–2, 4; homelands, 108, 191
baptisms, *65*
Basson, J. K., 54
Beckett, Caitlynn, 67
Bench Marks Foundation, 38, 108, 124, 138
Beningfield, Jennifer, 108
Berardo, Jose Manuel Rodrigues, 112
best practices, 30, 66, 82, 188, 198
biological, race as, 10
Bizos, George, 63
BKS, 79–80
Black Economic Empowerment, 114, 159, 191
Black history, 38, 124
Black intellectuals, 10–11, 217n54
Black Lives Matter (2016), 192, *193*
Black Panther Party, 26
Bloom, Jack, 151–52
Blyvooruitzicht village, 54
Bobbins, Kerri, 170–71
Boudia, Soraya, 28
Bourke-White, Margaret, 90–91
Brakpan superdump, 119, *121*, 125, 225n60
Breckenridge, Keith, 9, 34–35
Brenk report, 70–71, *71*, 72
bricks/brickmaking, 20–21, 184, 201
BS Associates, 69–70
buffer zones, 106–8, 113, 164, 166; around mine residue areas, 170, *171*, 182; Riverlea Extension, 124
Busby, Chris, 140, 142
Butcher, Siân, 105, 113, 165, 224n45

cancer, 78, 87, 126, 141; leukemia, 138; lung, 31, 42, 67, 157, *159*
capitalism, 10, 42, 103, 139, 165, 218n20; boom-and-bust, 41, 50, 59; democracy and, 192; growth and, 199–201, 235n5, 235n7; shell game, 158–59. *See also* racial capitalism
cemeteries, 40
Centre for Environmental Rights, 81, 138, 188, 190
Césaire, Aimé, 11
Chamber of Mines, 38, 53, 54, 56–57, 101; building exterior, *122–23*, 124; Vegetation Unit, 101, *102*, 103–4
Charlton, Sarah, 182
Chernobyl, 31, 32, 78, 130, 140
chiefs, 132, 191, 192
children, 153, 159, 192; playing on mine dumps, *146*, *147*, 148; radiation exposure, 139, 143; and school, 1–2, 60, 135, 136
Chirikure, Shadreck, 9, 35
City on a Hill, A (Phasha), 174–75, *176–78*, 178–79
city-region concept, 167–68. *See also* Gauteng City-Region (GCR)
class action lawsuits, 89
climate change, 6, 78, 199, 201; warming, 44, 45
coal, 22, 29, 53; coalfields, 61, 163; mines, 38, 88
Cole, Ernest, *91*, 91–93; *House of Bondage*, 4, *38*, *39*, 90, *94–96*, *107*
colonialism, 6, 13, 23, 41, 118, 218n20; geologists and, 37; molecular, 31
color blindness, 5, 15, 26, 200, 201
community consultation, 190–91, 192. *See also* public participation
conservationism, 13–14; protection of mine dumps, 118–19
constitutional rights, 14, 56, 72, 78–79, 137, 169
Cook, William, 101–4
copper, 29, 35, 159, 174, 175
corruption, 115, 175, 204, 235n13; RDP allocation and, 151, 154, 156
Council for Geosciences, 74, 141
Council on Scientific and Industrial Research (CSIR), 53–54, 195
Cradle of Humankind, 35, *36*, 64, 79

Crown Mines, 105–6, 107, 113, 119, 124

curies of rock, 42–43, 66

cyanide, 39, 48, 86, 100–101

democracy: capitalism and, 10, 192, 200–201; election of 1994 and, 2, 41, 104, 129; liberal, 4–5; Mandela's comments on, 60; participatory, 154; transition and dispensation of, 41, 55, 58, 165, 185, 192. *See also* constitutional rights

Democratic Alliance (DA), 136, 180, 181

demographics: of Gauteng Province, 27–28; of Krugersdorp and Kagiso, 133–34, 227n2; Wonderfonteinspruit population, 72; of WRDM, 79

Department of Environmental Affairs (DEA), 141, 152, 187, 188, *189*

Department of Mineral Resources (DMR), 69, 141, 152, 188, *189*, 193–94; inspectors, 189; jurisdictional battles, 187; Mintails and, 158–59, 160

Department of Water Affairs and Forestry, 59, 61, 72, 81, 82; investigation into water quality, 54–55, 75; permits, 119, 127

derelict mines. *See* abandoned mines

Diya, Mzawubalekwa, 89

Dlamini, Jacob, 14, 38

Donaldson Dam, *65*

drinking water, 57, 70, 83, 126; acceptable levels of uranium in, 76–77; access to clean, 14, 60, 71–72, 74

drug use, 135

dumps. *See* mine dumps; sand dumps; tailings piles

Durban Roodeport Deep (DRD) Gold, 113, 117–19, 125, 153

dust, 7, 25, *115*, *182*; clouds, 97; control, 101–3; dump removal and, 148; inhalation and diseases, 32, 39, 87–89, 124, 126, 140; proliferation of, 85–86; ventilation and, 143, 146

du Toit, J. Stephan, 73–75, 141, 146

Dying for Gold (2018), 89, 223n8

Earthlife Africa, 71, 127, 138

East Rand, 40, 102, 125; Jerusalem informal settlement, *100*, *147*

East Rand Gold and Uranium Company (ERGO), 113, 119, *185*, 186, 225n60

economization of life, 56

education system, 1–2, 4

electricity supply, 163

Enlightenment humanism, 5, 10, 199, 200

environmental impact assessment (EIA), 82, 126, 149, 170, 181, 188–90

environmentalism, 11, 61; Black exclusion from, 13–14; health and, 16; jobs and, 12–13, 15; West Rand, 79–80; women's participation in, 136. *See also* Liefferink, Mariette

environmental justice, 8, 13, 14, 27, 31, 137, 170

Environmental Management Frameworks (EMFs), 79–80

eucalyptus, 101

evictions. *See* relocations, forced

externalities, 29–30

"Fancies Idle" (Abrahams), 93, *98*

farmers, 48, 72, 74; complaints about water quality, 53–55, 64; leukemia rates among, 138; protest against superdump, 125–27

Federation for a Sustainable Environment (FSE), 72, 127, 137, 157; establishment of, 63–64; suit against Mintails, 160; Tudor Shaft activism, *146*, 148, 152

Ferdinand, Malcom, 11

fish, 52, 70, *71*, 140, 163

Fisher, Mark, 235n7

food chain, 57, 64, 74, 163

foreign investment, 9, 114

formal and informal practices, 201–3; economies, 174, 182, 235n9; mining, 22–23, *24*. *See also* informal settlements

Freedom Day, 129

Freiburg, Germany, 66–68

funding: AMD remediation, 169; "European green funds," 160; growth vs. repair, 30; mine rehabilitation, 75, 158–60, 189, 194

Gaule, Sally, 92; *Stardust* series, 119, *120–23*, 124
Gauteng City-Region (GCR), 167–70, *171*, 183, 206
Gauteng City-Region Observatory (GCRO), 168–70, *173*, 174, *180*
Gauteng Province, 16, 31–32, 61, 118, 125, 164; administrative regions, 23, 167–68; EMF, 80; housing, 179–81, *180*; mine residue areas, 28, 166, *167*, 169; platinum belt, 190; population, 27–28; radiation exposure in, 68; Strategic Water Management Plan, 59; uranium infiltration, 44, 67; water supply, 47–49
Gauteng's Department of Agricultural and Rural Development (GDARD), 166
geological time, 34–35, 43, 44. *See also* Anthropocene
Gevisser, Mark, 41, 117, 124
Gilmore, Ruth Wilson, 10
global economy, 168
Godfrey, Ilan, *24*, *125*
gold: deposits or layers, 34–35, 37; grades, 29; prices, 104, 106, 109, 112, 113; relationship to uranium, 42
Goldblatt, David, 93
Gold Fields, 53, 54, 157
gold mining, 9, 20, 85, 105; early evidence of, 37; grinding and extraction, 22, *24*, 39, 48, 112–13, 127; reprocessing, 113–14, 117; total mined, 42. *See also* Crown Mines; Durban Roodeport Deep (DRD) Gold; Harmony Gold
Gold Reef City Theme Park, 20, 64
Gordimer, Nadine, 93, 96–97
grassing, 101–4, *102*, 109
Great Zimbabwe, 35, 37
Green, Lesley, 14
greenwashing, 119, 199, 201
Grootvlei mine, 53

growth rhetoric, 30, 199–201, 235n5
Guinier, Lani, 31

Haraway, Donna, 16
Harmony Gold, 71, 104, 114, 187–88
heavy metals, 54, 74; in acidic water, 25, 49, *50*, 53, 137; in vegetables, 97, 100
Hemson, David, 10
Hendricks, Lindiwe, 72
Hepler-Smith, Evan, 78
heritage, 66, 118–19, 124, 132. *See also* Cradle of Humankind
Hinton, Elizabeth, 213n7
Hoover, Herbert and Lou, 52
hostels, 130, 133, 134
housing, 13, 14, 21, 68, 104; backlog, 79–80, 179; backyarding, 184, 201; construction materials, 201–2; forced relocation and resettlement, 108–9, 132–33, 142; megaprojects, 179–81, *180*; shacks, 130, 142, 149, *150*, 153–54, 183; supply and demand, 106, 165–66; Tudor Shaft, 153–54, 156; ventilation, 143, 146; wind direction and, 86, 97. *See also* Reconstruction and Development Programme (RDP) homes
Hudson, Peter, 11
human evolution, 35
Humby, Tracy-Lynn, 14, 148, 188

identity, 21, 92
ignorance, 6, 29, 30, 44, 54; epistemologies of, 15–16, 169; manufactured, 16, 69, 71, 146, 198
Iheka, Cajetan, 8
imbizo/izimbizo, 154, 156
imperialism, 11, 107
Industrial Development Corporation (IDC), 114
infill, 179, 183–84, 186
informal settlements, 15, 49, 72, 74, 81, 170; Bekkersdal, *65*; construction materials, 201–2; on/near mine dumps, 44, 97, *100*, *147*; total WRDM residents in, 79. *See also* Tudor Shaft
infrastructure, 5, 9, 13, 16, 109; apart-

heid's, 169–70; housing, 165–66; justice and, 206; lack of basic, 14, 60–61, 134–35; racial capitalism and, 11, 12, 23, 205; regulation of, 55–56; road, 106; separatist ideologies and, 41; urban planning, 25–26, 106; water and electricity, 48, 53
Institute for Water Quality Studies (IWQS), 57–58
International Atomic Energy Agency (IAEA), 77–78
International Commission on Radiological Protection (ICRP), 56, 57, *131*, 146, 148, 152, 202
International Network for Acid Prevention, 66
iron, 35, 37, 39, 51
isiZulu, 22, 133, 154

Jenkins, Destin, 11
jobs/environment dichotomy, 12–13, 15
Johannesburg, 40, 47, 48, 64, 124, 175; description, 183; emigration to, 22; housing development, 181; land-ownership, 105; mine dumps, 86, 93, *110–11*, *115*, 174, *177*; Sandton, 59–61; Top Star drive-in, 115–19, *120–21*, 125; town planning, 100, 116, 164; view from space, 19, *20*; water source, 76
Johnson, Willard R., 42

Kagiso township, 7, 131, 194, 227n2; Community Work Program (CWP), 134–36; history, 132–34, 227n5; radon concentrations, 156–57, *158*
Keeling, Arn, 67
Khoda, Moekhti, 192, 200
Khumalo, Vusumuzi, 132–33, 227n5
knowledge, 7–8, 15, 16, 78, 202–3; territorial, 105, 113
Kongiwe Environmental, 233n68
Kruger National Park, 14, 38
Krugerrands, 42
Krugersdorp, 112, 132, 133–34, 227n2
Krugersdorp Game Reserve, *59*, 64, 74

labor. *See* migrant labor; mine workers
land rights, 190, 191, 192
languages, 8, 20–21, 133, 227n2; apartheid regime and, 1–2
lead, 43, 56, 74–75, 100
Legassick, Martin, 10
legislation: mining, 170, 186–89; post-apartheid, 191; water pollution, 53, 101; workplace, 86
le Roux, Hannah, 106
Lesotho, 22, 48–49, 76, 87, 223n8
liability, 59, 82, 104; Mintails', 158; perpetual, 103, 187–88
Lichtenstein, Alex, 93, 96
Liefferink, Mariette, 72, 74–75, 78, 126, 137; Amatshe Mining and, 160–61; creation of FSE, 63–64; Mintails and, 148–49, 153, 159–60; NNR and, 139, 194–95; on public participation, 80; toxic tours, 157–58, 230n67; victory over Shell Oil, 61, 63; work approach, 81, 138–39
Life magazine, 90
Livingston, Julie, 13
load shedding, 175
Loots, Cobus, 160
Lunuberg, Justice, 126

Madida, Sipokazi, 118
Mahatammoho cooperative, 126
Mahlangu, Dumisani, *24*
Main Reef Road, 19–20, 23, 25, 100; Riverlea community, 124, *125*, 181
Makgoka, T. M., 187
Makhura, David, 180, 181
Malan, Daniel, 112
Mandela, Nelson, 10, 60–61, 63, 129
Mandlo, Daniel, *24*
Marikana mine, 191, 192
Masekela, Hugh, 38–39
Matakoma Heritage Consultants, 118
Mavhunga, Chakanetsa, 9, 14
Maviyane-Davies, Chaz, 9; montages, *18*, *46*, *84*, *128*, *162*, *196*
Mbembe, Achille, 10–11, 16–17, 164, 217n59

McCarthy, Terrence, 169–70
McCulloch, Jock, 87
medical examinations, 87, 89, 92
mercury, 22, 49, 56, 74, 100
Merriespruit, 104–5
meteor strikes, 34–35
migrant labor, 38–39, 48, 79, 92, 96, 133
Mills, Charles, 4–5, 13–14, 15–16, 199
mine dumps: as buffer zones, 107–8; compared to tombs and pyramids, 93, 96, *98, 99*; contents and proliferation of, 85–86, 101; housing and, 97, *100*, 181, 182, 201; Johannesburg, 86, 93, *110–11, 115*, 174, *177*; within the mining belt, *185*; Mooifontein (Riverlea), 124, *125*; protection of, 118–19; reclamation of, 115–19, *120–21, 185*, 186; rehabilitation of, 126, 160–61, 164, 166, *167*, 193–94; remining/reprocessing of, *34*, 112–15, *115*, 225n60; removal of, 148–49, 202; residual governance and, 100; Soweto, *26, 182*; spraying of, 101; vegetation for stabilizing, 100–104, *102*, 109, 116, 126, 185. *See also* sand dumps; scrap materials; tailings piles
mine lands, 107, 109, 166, 184, 186, 195; as land mines, 164, 182
mine residue areas, 16, *63*, 164; Gauteng's, 28, 166, *167*, 169, 170, *171*
mine shafts, 85, 86, 103, 106, 159; acid water and, 45, 49–50, 109; collapse of, 22, 25, 39; pumping water out of, 40, 48, 49, 169
Mine Waste Solutions, 125–26, 127
mine workers: Cole's images of, 90–93, *94–96*; health risks and diseases, 44, 51, 86, 87–88; informal, 22, *24*, 202, 207; injury and compensation, 89, 93, *96*; recruitment, 22, 38–39; shaft-sinkers, 48; treated as waste, 31; unions, 88, 191; violence against, 23; working conditions, 39–40. See also *zama zamas*
mining belt, 183–85, *185*

Mining Charter (2004), 190–92, 193
mining industry, 9, 10, 31, 109; active sites worldwide, 199, *200*; community consulting, 190–91; extraction waste data, 198–99, *200*; hand tools and mechanization, 39–40; inspectors, 189; knowledge monopoly, 165; magazines, 90; manufactured ignorance of, 69; public-private partnerships, 185–86; racial discrimination, 87–88; regulations and legislation, 55–57, 186–88, 202; remediation consulting firms, 164, 166, 189–90, 231n10, 233n66, 233n68; sand and dust generated by, 7, 85–86; violence of, 23, 25. *See also* abandoned mines
mining rights, 159, 160, 191
Mintails, 113, 152–53, *185*, 186; environmental liability and rehabilitation, 158–60; mine dump removal, 148–49, 202; Princess Pit, 151
Misapitso, Lucas, 154
Mjadu, Phumla Patience, 134, *135*, 149, 153
Mkhize, Boyce, 142
Mnwana, Sonwabile, 191
Mogale City Local Municipality, 73, 75, 131, 148, 152–53; evictions and relocations, 141–42; history of, 132–34; lung cancer deaths, 157, *159*
Mokoena, Lydia, *65*
molecular bureaucracy, 78, 137
Mondi, Lumkile, 114
Moodley, John, 181
Morojele, Mpethi, 124
Moshupya, Paballo, 156
Mpofu-Walsh, Sipho, 41, 69, 191
Mpumalanga, 61, 82, 163, 188
Munro, A. H., 56
Murphy, Michelle, 56

National Environmental Management Act (1998), 79, 186, 188
National Heritage Resources Act (1999), 118

National Imbizo Focus Week, 154, 156
National Nuclear Regulator (NNR),
    57, 74, 194–95; deal with Mintails,
    148–49; Mogale City evacuations
    and, 141–42; radiological risk assess-
    ments, 69, 71, 72, 139–40, 142–43,
    146, 152
National Party, 9, 106, 112
National Union of Mineworkers (NUM),
    88, 191
National Water Act (1998), 61, 79, 186
Ncwana, David, 140
Ndlovu, Linda, *24*
new energy systems, 29
Nodal Review Policy, 183, 184
Ntsepo, Clara, *150*, 151
Nuclear Energy Act, 56
nuclear power, 29, 31, 44, 63
nuclear weapons, 43–44, 52
nyaope, 135

Oatway, James, *143*
Oesi, Joseph, 192, *193*
oligarchs, 114–15, 204
One Environmental System, 188, *189*,
    193
On the Mines (Goldblatt and Gordimer),
    93, 96–97
Oppenheimer, Ernest, 40, 107
Optima magazine, 93
Orenstein, Alexander, 87
organized labor, 37, 88, 191
Oxpeckers Center for Investigative En-
    vironmental Journalism, 158

Padda Dam, 70
Pamodzi Resources, 114, 187
Pan African Resources, 160
passbooks (*dompas*), 38, 40
personhood, 92, 206
Phasha, Potšišo, 198; *A City on a Hill*,
    174–75, *176–78*, 178–79; "Scavenger
    Economies of the Mine Dumps," 171,
    172–73, 174
Pheto, David, 153, 156, *157*

pH levels, 51, 82–83, 102
Pieterson, Hector, 2
Plaatje, Sol, 88
Plan Associates, 184–85
planetary futures, 6, 7, 201; Anthropo-
    cene and, 11–12, 16–17
platinum mining, 103–4, 186, 190, 192
poetry, 98, 175, 178–79
Pofadder, 138
poverty, 60, 61, 80, 160; in Kagiso town-
    ship, 134–35; pollution and, 13, 15
Promotion of Access to Information Act,
    33, 170
property rights, 105, 165
Provincial Heritage Resources Authority
    of Gauteng (PHRAG), 118–19
public participation, 80, 129, 154, 170
public-private partnerships, 82, 180–81,
    185–86
pumping stations, 48, 49, 64, 159

quality-of-life surveys, 169, 170
quartzite, 21, 37

race riots, 2, 213n7
racial capitalism, 13, 27, 203, 206–7;
    Anthropocene and, 5, 12, 31, 32, 207;
    descriptions of, 10–11; residual gov-
    ernance and, 6, 82, 88, 93, 186, 198,
    204–5; sediments of, 23; violence of,
    50, 64
racial contract, 9, 41, 186, 198, 205,
    206; Mills's theory, 4–5, 15–16, 199
racism, 10, 16, 23, 40; archaeologists
    and, 37; ecocide and, 12, 25; mine
    workers and, 87–88; spatialized, 86,
    107; systemic and epistemic, 2, 4,
    5, 11
Radebe, Jeff, 154, 156
radiation exposure, 31, 42; breast milk
    and, 70; children vs. adults, 139, 143;
    ICRP recommended limits, 56, 57, 146,
    148, 202; IWQS study on, 57–58; NNR
    risk assessments, 69–71, 139–40,
    142–43

radioactivity: around West Rand Gold
Mine, *73*; contamination at Tudor
Shaft, 130, *131*, 139, 141–42, 149, 152,
160; of dust and sand, 7, 25; locations
of water bodies and, 170, *171*; of mine
water, 54–55, 57–58; pathways, 68,
70, *71*; of quartzite, 21; of uranium,
42–44, 66
radon gas: in construction materials,
201–2; lung disease from, 39, 42–43;
measuring, 143, 146, 152, 156–57,
158; in US homes, 44, 202
Rainbow Nation, 61
Ramaphosa, Cyril, 88, 191
Ramoruti, Jeffrey, 130, *150*, 151, 153,
228n43; photos of, *155*, *157*
Rand (Witwatersrand), 39–40, 52, 115,
130; American mining interests, 9;
description, 20; dust clouds, 97; hollow, 49, 51, 67, 85, 106, 164, 182;
inside-out, 85–87, 164, 170; uranium,
43, 68, 137; water source, 7, 47–48;
Western and Central Basins, 81–82.
See also East Rand; West Rand
Rand Mines Properties (RMP), 106, 108,
113, 165, 184
Rand Water Board, 48, 53, 168
Reconstruction and Development
Programme (RDP) homes, 151, 166,
181, 201; Soul City and Tudor Shaft
allocations, 153–54, 156; total built,
130
Red Ant Security Relocation and Eviction Services, 142, *143*, *144–45*,
149
regulatory standards, 13, 30, 82, 189,
198; for air pollution, 101; for radiation exposure, 56–57, 146, 148, 202;
for treating AMD, 82–83; for uranium
in drinking water, 76–78
Reichardt, Markus, 103, 104
religions, 40
relocations, forced, 40, 106, *107*, 130; of
Kagiso residents, 132–33, *150*; by Red
Ant brigades, 142, *143*, *144–45*, 149;

of Tudor Shaft residents, 141–42,
148–49
residual governance, 7, 16, 136, 182,
202, 207; apartheid's reliance on, 138;
artists and, 174; concept and dynamics, 6, 28–32; corruption and, 204;
democratic dispensation and, 55–56;
invisibility and, 8; manufactured ignorance and, 69; of mine dumps, 86,
97, 100; minimalism of, 32–33, 49, 55,
66, 103, 186; mining companies and,
191–92; politics of, 165; proponents
of, 130, 139; racial capitalism and, 6,
82, 88, 93, 186, 198; struggles against,
126, 131, 161, 164, 166, 205–6
residues, term usage, 28–30, 218n13.
See also dust; mine residue areas
resource extraction value, 27
Rio Tinto, 51
Rittel, Horst W. J., 25–26
Riverlea community, 108, 124, *125*, 181
Robinson, Cedric, 10
Robinson Dam, 201
rock removal, 85
Rubin, Margot, 180, 181

sand dumps, 53, 85–86, 97, 102, *131*.
See also Brakpan superdump
sangoma (traditional healer), *65*
sanitation services, 14, 40, 55, 134
satellite imagery, 19–20, *20*
scale, 13, 29; economies of, 205; geographic, 168, 169; geological time,
34–35; governance, 131; industrial,
21, *34*; interscalar dynamics, 33, 201,
202; planetary, 11–12; spatial, 19–20,
33, 106
"Scavenger Economies of the Mine
Dumps" (Phasha), 171, *172–73*, 174
science and technology studies (STS),
16, 216n18
scrap materials, 136, 158, 202; scavenging for, 171, *172–73*, 174–75,
176–78
sediments: of racial capitalism, 23;

radioactivity in, 57, 70, *71*, 72; sea, 34; uranium concentrations in, 58, 76

segregation, 10, 23, 106–7, 205; cemeteries and, 40; drive-in theaters and, 116

Selaole, Susan, 149, 151–52

Sello, Kefiloe, 48

service delivery protests, 14

shafts. *See* mine shafts

Shell Oil, 61, 63

Sibanda, Calvin, *24*

Sibanda, Precious, *24*

Sibanye-Stillwater, 89, 157

Sibanyoni, Jonas, 195

Siddiqi, Asif, 200

silicosis, 31, 86, 87, 89

sinkholes, 48, 53, 54, 68

Sinqobile township, 151, 228n43

Sisulu, Lindiwe, 179

slimes dams, 53, 86, 101–2, 103–4, 114, 134; in the mining belt, *185*; runoff from, 68, 70

slurry, 86, 103–4, 127

Smit, Susanna, 126

Smuts, Jan, 109

social contract, 4–5, 13, 15, 199

Socio-Economic Rights Institute (SERI), 137, 148, 152, 153, 156

soil contamination, 97, 152, *155*, 203, 229n63; maize crops and, 140, 153

solutionism, 27, 103–4, 201; simplification and, 16, 31, 82, 205

South African Human Rights Commission (SAHRC), 194–95

Soweto, 19, 108, 164, *182*; Klipspruit district, 40; Lufhereng housing megaproject, 181–82; student uprising (1976), 1–2, *3*, 133, 213n7; suburbs, *26*; urban planning, 184

Spain, 51

Spatial Development Framework 2040 (SDF 2040), 183–84

spatial injustice, 164–65, 167, 179

Spatial Planning and Land Use Management Act (2015), 183

squatter movements, 106

state capture, 114–15, 191, 204, 205

Steffen, Robertson and Kirsten (SRK), 233n66

"Stimela" (Masekela), 38–39

stokvel (mutual aid schemes), 136

Strategic Area Framework (SAF), 184–85, *185*

student movements, 1–2, *3*, 133

suburbs, *26*, 59, 108, 124, 170; upwind, 25, 106

suffering, image of, 8

sulfates, 12, 53, 74, 83, 201

sulfide minerals, 66

superdumps, 119, *121*, 125–26

super wicked problems, 31, 76, 115, 183, 186, 194, 201; criteria for, 27

"sustainable mining," 195

tailings dams, 28, 97, 100, 101, 142; failures, 25, 103–5

tailings piles, 29, 41, 115–16, 152; along Main Reef Road, 20, 23, 25, 100, 124; compared to pyramids, 32, 96; temporality of, 97; uranium and, 43, 44; vegetation for stabilizing, 100–101. *See also* mine dumps

technopolitics, 30, 41, 55, 197, 207, 213n14; environmental, 14; of the racial contract, 5, 186, 206

Toens, Dennis, 137–38

Top Star drive-in theater, 115–19, *120–21*, 125

Torres, Gerald, 31

toyi-toyi, 14, 32, 149

traditional governance, 191, 233n70

Truth and Reconciliation, 118, 204

Tshiamiso Trust, 89

Tudor Shaft, 7, 134, *146*, 194; forced relocations, 141–42, 149; mine dump removal, 148–49; politics and community struggles, 151–52, *155*; radioactive contamination, 130–31, *131*, 139, 160; radon levels, 156; RDP housing, 153–54, 156

UN Security Council, 2, 213n6
uranium, 42–44, 83, 201, 218n20; deposits or layers, 34–35; extraction plants, 54, 112–14, 127; Freiburg conference on, 66–68, 221n52; leukemia link, 138; limit in drinking water, 76–77; prices, 113; scientific studies, 137–38; in stream water, 56, 58, 69, 74, 75–76; value and significance of, 109, 112
urban planners, 25–26, 40, *177*; activists, 164–65; buffer zones and, 106–8, 116; mining belt and, 183–85, *185*
US Environmental Protection Agency, 66

Vaal River, 53, 125, 127
van der Merme, Charles, 181
vegetable gardens, 97, 100, 135, 140, 153
vegetation, dump, 100–104, *102*, 109, 116, 126, 185
Vergès, Françoise, 10, 12
Verwoerd, Hendrik, 106, 181
Villiers, Dawid de, 152
violence, 8, 23, 174; evictions and, *143*; in Johannesburg, 183; in Kagiso township, 135; protests and police, 2, 133, 151, 152, 191; of racial capitalism, 11, 20, 50, 64; slow, 25, 49, 52, 103, 109; volumetric, 7, 130, 166, 183, 199
voids, mine, 25, 48, 49, 59, 64, 68, 82
voting, 129, 205

Waggitt, Peter, 67, 221n56
waiting, forms of, 92, *94–95*, 136, 227n13
waste: Anthropocene and, 27, 28; extraction, 198–99, *200*; histories, 5; nuclear, 44; people/places treated as, 6, 31, 32, 131, 135, 198, 207; race and, 12, 13–14, 41; rock, 40; scale of, 29; South Africa's total production, 28; work, 135–36, 227n15
Water Act (1956), 53, 101
water management, 59, 79, 195; quality studies, 53–55, 57–58; treatment plants, 82–83, *83*, 159, 170
water pollution, 7, 25, 40, 127; acidification, 45, 49–51, 59, 109; assessments and reports, 55, 69–72; radioactive tailings and, 170, *171*; radium levels, 54–55; river effluents, 48, 53; uranium contamination, 56, 58, 67–68, 69, 76–77; in Wonderfonteinspruit catchment, 56, 63, 72, 74–76
Water Research Commission (WRC), 58, 59, 69, 81
water sources, 7, 47–49, 53, 54, 81; protests over access to, 60–61, *62*, 71–72; scarcity, 23, 83; West Rand municipalities, 76. *See also* drinking water
Webber, Melvin, 25–26
Wendel, G., 42–43, 66
West Rand, 40, 42, 57, 67, 113, 138; acid water decant, *50*, *59*, 61, 63; environmental management, 79–80, 114; lung cancer deaths, *159*; mine dumps, *34*, 125; radiation levels, *73*; Robinson Lake, *203*; uranium contamination, 56, 68; volume of void, 49; water neutralization plant, 82–83, *83*, 158; water quality, 53, 64; water source, 76
West Rand Consolidated Mines, 112, 130
West Rand District Municipality (WRDM), 79–80, 131
wetlands, 68
whiteness, 41, 93; privilege of, 106; women and, 21–23, 61, 90, 97
white settlers, 40–41, 100
white supremacy, 4, 5, 11, 132, 200
wicked problems, 25–27, 28, 32. *See also* super wicked problems
Wilge River, 163
Winde, Frank, 67–68, 75–76, 77, 138, 140
Witwatersrand Native Labor Association (WNLA), 38, 93
Witwatersrand plateau. *See* Rand (Witwatersrand)

WoMIN, 138

Wonderfonteinspruit catchment, 194–95; heavy metal levels, 74; radiological contamination, 72–73, *73*; radiological risk assessment, 69–71; uranium contamination and remediation, 56, 63, 69, 75–76

Woodson, Carter Godwin, 4

World Bank, 9, 48, 165

World Health Organization (WHO), 76, 77

World Summit on Sustainable Development (2002), 59–61, *60*, *62*, 66

Wymer, Denis, 57

zama zamas, 159, 174, 202, *203*, 207, 217n2; meaning, 22

Zondo, Raymond, 204

Zuma, Jacob, 114, 191, 204